HOW
OLD
IS
THE
UNIVERSE?

DAVID A. WEINTRAUB

HOW
OLD
IS
THE
UNIVERSE?

PRINCETON UNIVERSITY PRESS   PRINCETON AND OXFORD

Library of Congress Cataloging-in-Publication Data

Weintraub, David A. (David Andrew), 1958–
How old is the universe? / David A. Weintraub.
p. cm.
Includes index.
ISBN 978-0-691-14731-4 (hardback : alk. paper)   1. Solar
System—Age.   2. Earth—Age.   3. Cosmology.   I. Title.
QB501.W45 2011
523.1—dc22   2010009117

British Library Cataloging-in-Publication Data is available

This book has been composed in Goudy

Printed on acid-free paper. ∞

Printed in the United States of America

1 3 5 7 9 10 8 6 4 2

To

Mark and Judith

and

Erica and Bruce

∴

∵ CONTENTS ∵

HOW
OLD
IS
THE
UNIVERSE?

···CHAPTER 1···

Introduction
13.7 Billion Years

I do not feel obliged to believe that that same God who has endowed us with senses, reason, and intellect has intended to forgo their use. . . . He would not require us to deny sense and reason in physical matters which are set before our eyes and mind by direct experience or necessary demonstrations.

—Galileo Galilei, in "Letter to Madame Christina" (1615), translation by Stillman Drake, *Discoveries and Opinions of Galileo* (1957)

Astronomers are at an enormous disadvantage, compared with other scientists. A biologist can bring a collection of fruit flies into his laboratory, encourage a particular behavior among those flies, and apply all the tools of his trade to studying that behavior. A chemist can mix chemicals together, heat them up or cool them down, and study how they react in the controlled environment of her laboratory. A geologist can hike up a mountain, collect rocks from a particular outcrop, and return these samples to his laboratory for analysis. A physicist can power up a laser and test the mechanical properties of a newly-created polymer and can do this on her vibration-isolated, experimental table. Astronomers? They cannot drag the stars into their laboratories. They cannot make stars hotter or cooler to see how they behave when their temperatures change. They cannot slice open galaxies in order to peer into their cores. Astronomers can only take what the universe offers—light and a few small rocks—and make the most of it.

For centuries, astronomers have measured the brightnesses, colors, and positions of objects in the nighttime sky, as one generation after another has sought to understand the nature and behavior of the remote objects shining in the heavens. Using basic principles of geometry, and the physics describing light, heat, and gravity, astronomers deduced that some of those glimmering objects in the sky are akin to our Sun: They are stars. They also discovered that stars have a wide range of sizes, masses, and temperatures, and that stars are born, live out their lives, and then die. Proving seemingly obvious things, however, like the fact that stars are distant (which prompted the thorny question, how distant?), was extremely difficult. To answer the question of how far away stars are—not to mention how hot, how massive, and how large or small they are—astronomers had to learn how to take detailed measurements of celestial objects that lie at great distances from our telescopes. Since the stars could not be brought to Earth to be weighed and measured, astronomers had first to develop the right tools for measuring stars. Then they were able to apply their knowledge of concepts like Doppler shifts, radioactivity, and nuclear fusion to the measured properties of stars, and answers to all sorts of previously unanswerable questions, including ones that had frustrated astronomers since antiquity, began to rain down from the heavens. This deluge of evidence led eventually to an astonishing and hard-won intellectual triumph on the grandest possible scale: the answer to one of the most fundamental questions ever to puzzle humanity, How old is the universe?

Astronomers have made great strides. Barely 200 years ago, they were unable to measure the distances to even the closest stars, let alone use the properties of stars to measure distances to galaxies 100 million light-years away. Now they boldly claim to know the age of our universe to an accuracy of better than 1 percent: 13.7 billion years. You now know the answer to the question on the cover of this book, and so it is clear that the book you are holding in your hands is not a mystery. But, it is *about* a mystery. How have 400 years of science brought us to this point at which astronomers, cosmologists, and physicists can claim that the universe came into existence at a specific moment, 13.7 billion years ago? And how much confidence should you have in this statement?

Ask any astronomer why she believes the universe is 13.7 billion years old, and she will tell you that she does not *believe* that the universe is 13.7 billion years; she *knows* that it is 13.7 billion years old—give or take a hundred million years. Why are astronomers so confident? It turns out that their certainty is not hubris. They know that this number is the only valid answer to the question of the age of the universe because it emerges from a meticulous interpretation of all the data—from rocks, stars, galaxies, the whole universe—that humanity has painstakingly gathered over the centuries. It is the only answer that is consistent with the laws of physics as we know them and with the firm logic of mathematics and justified by the collective labors of astronomers, as well as of chemists, mathematicians, geologists, and physicists. The answer rests, in fact, on very solid foundations.

But why *exactly* do twenty-first century astronomers think that 13.7 billion years is the right answer? Why not 20 billion years? Why not 6,000 or 50 million or 1,000 trillion years? How do astronomers know that the universe even has an age— that it is not eternal?

A thorough and persuasive answer to the question of the age of the universe requires that we follow in the footsteps of Galileo and of the many other curious scientists who have come before us, and that we explore the great breadth of knowledge that lies at the core of modern astronomy. The answer to our title question is derived from evidence gleaned from many areas of inquiry that fits together, much like foundation stones in the base of a solid wall. Science in general progresses when curious people pose smart questions and then answer them, or identify problems (the stones in a foundation that do not fit well and consequently make the wall unstable) and then solve them, thereby making the entire scientific structure sturdier and more reliable. The astronomers and other scientists who have made some of the greatest achievements leading to our discovery of the age of the universe will appear in these pages. Some of them may be utterly unfamiliar to most readers, while others bear names that have become iconic. They include William Herschel, Joseph Fraunhofer, Friedrich Wilhelm Bessel, Edward Charles Pickering, Annie Jump Cannon, Henrietta Leavitt, Ejnar Hertzsprung, Henry Norris Russell, Vesto Slipher, Harlow

Shapley, Edwin Hubble, Fritz Zwicky, George Gamow, Walter Baade, Vera Rubin, Arno Penzias, Robert Wilson, Robert Dicke, and James Peebles.

These scientists, and many others, advanced our knowledge of the universe through their work, often correcting errors in our collective wisdom and enabling the entire field of astronomy to leap forward. In these pages, you will learn about these discoveries and their significance. By walking in the steps of the astronomers who ascertained the ages of certain objects in the universe and hence deduced the age of the universe itself, you will learn just how solid are the foundations of our astronomical knowledge. When you have finished this book, you will understand the central claims that strongly support the conclusion that the universe is just under 14 billion years old:

- The oldest known meteorites in the solar system are 4.56 billion years old. From all of our knowledge about how stars, planets, and asteroids—the parent bodies of meteorites—form, astronomers are confident that the Sun and all the other objects in our solar system were born at nearly the same time as these meteorites. This age for the Sun is consistent with all observations and theorists' understanding of the physics of the Sun, of other stars, and of the life cycle of stars. As the universe must be older than every object contained in it, including our solar system, the universe clearly must be at least 4.56 billion years old.
- The oldest white-dwarf stars in our own Milky Way Galaxy have been cooling off (as white dwarfs) for about 12.7 billion years. Since the white dwarfs formed from stars that lived and then died, and since those stars that died had lifetimes of a few hundred million years, this age requires that the Milky Way, and certainly the universe, must be older than about 13 billion years.
- The oldest globular clusters in the Milky Way have measured ages of about 13.4 billion years. Therefore, the Milky Way and the universe itself must be at least marginally older than these oldest globular clusters.
- Cepheid variable stars in galaxies out to distances of 30 megaparsecs (100 million light-years) trace the expansion of the universe. They

permit astronomers to calculate how long the universe has been expanding at the current rate in order for the galaxies to have achieved their current separation distances. Assuming that the expansion rate of the universe has been constant over the history of the universe, which all evidence suggests is very close to a correct assumption, the age of the universe is just a bit greater than 13.5 billion years.

- Maps of the cosmic microwave background radiation contain information about the range of temperatures and the sizes of the structures in the universe when this radiation was emitted. When combined with information about dark matter and dark energy and the rate of expansion of the universe, the most thorough and rigorous analysis of the maps of the cosmic microwave background reveals that the age of the universe is close to 13.7 billion years.

These claims only make sense if one knows what white dwarfs, globular clusters, Cepheid variable stars, and the cosmic microwave background are. Our story, an intellectual voyage of discovery spanning centuries, will be interwoven with the science that elucidates the nature of these celestial objects and explains how astronomers have learned their ages, and how, in turn, they can claim to know the precise age of our universe. But, where to begin? As it turns out, the search to uncover the age of the universe started right here at home, with the mystery of the age of the Earth.

·:1:·

THE AGE OF OBJECTS IN OUR SOLAR SYSTEM

··· CHAPTER 2 ···

4004 BCE

In the beginning, God created Heaven and Earth, *Gen. I.* V.I. Which beginning of time, according to our chronologie, fell upon the entrance of the night preceding the twenty third day of October in the year of the Julian calendar, 710.

—James Ussher, in *The Annals of the World* (1650–1654)

How old is the Earth? Clearly it cannot be older than the rest of the universe, so if we could determine the age of the Earth we would have a minimum age for the universe. And that would be an excellent starting point for investigating its total age.

We live on the Earth. By virtue of that location, we are able to learn more about the Earth than about other places in our solar system, let alone our universe. So let us begin attempting to measure the age of the Earth by observing the world around us. From this beginning, we can then use the tools of astronomy to look outward, taking measurements to determine the age of celestial objects and, eventually, of the entire universe. How exactly do we begin?

Aristotle's Eternal Heaven

Without the tools of modern science, what process would a scholar follow in order to determine the age of the Earth?

For 2,000 years, Aristotle provided an answer to that question that satisfied most seekers. In his treatise *On the Heavens*, which he wrote in about 350 BCE, he declared "there is one heaven only, and [that] it is ungenerated and eternal." The universe, then, has existed for all times past and will continue to exist into the infinitely remote future. The wheeling sphere of the stars, the "one heaven," is eternal. Aristotle's claim for a universe without beginning and without end rested on his ideas for physics and for motion: the celestial sphere moves in circular motion around the Earth, circular motion is the perfect motion (in contrast to up/down or sideways movements), objects moving in circles have achieved their ultimate purpose, their *telos*, and have neither the ability nor any reason whatsoever to slow down, speed up, or change direction. Objects cannot become perfect since they would have to change from imperfect to perfect, which is not possible, according to Aristotle; therefore, the heavens have always been perfect and must have existed in their current state for all of eternity. Aristotle then adds arguments to prove that the Earth, too, is eternal: a sphere cannot exist without a center, and the Earth is very clearly the center of the heavenly sphere; therefore, since the heavenly sphere is eternal, the Earth also must be eternal. Aristotle's logic and reasoning were elegant, sophisticated, powerful, and regrettably also wrong. Nevertheless, despite obvious flaws in his metaphysical arguments, Aristotelian natural philosophy dominated human understanding of the physical universe until the seventeenth century.

The revolutionary ideas that finally overthrew Aristotle's cosmology emerged with the publication of Nicholas Copernicus' monumental work *The Revolutions of the Celestial Orbs* in 1543 and culminated with the publication of Isaac Newton's *Principia* in 1687. During the century and a half between the appearances of these two great books, new astronomical ideas emerged from the work of, among others, Tycho Brahe, Michael Maestlin, Johannes Kepler, and Galileo (who built his first telescope in 1609). By the early seventeenth century, natural philosophers were beginning to seriously question Aristotelian cosmology. Perhaps the universe is not eternal. If the universe is not eternal, it had a beginning. If the universe had a beginning, the Earth also had a beginning. Almost overnight, the age of the Earth emerged as an important topic for scholarship. Copernican cosmology, however, unlike Aristotelian cosmology, offers no answers. The heavens, as understood by the Copernicans, are mute as to their age. What, then, is the age of the

Earth? And through what methodology would seventeenth-century scholars have even begun to try to answer this question?

Biblical Chronology

In the seventeenth century, Bishop James Ussher calculated the length of time that had lapsed since Adam appeared in the Garden of Eden. He reasoned as follows. The words of Chapter 1 of *Genesis* detail how Adam appeared on the sixth day after the beginning, after "God created the heavens and the earth." Therefore, according to the logic used by Bishop Ussher, the period of time since "God created man in his image" plus five days provides an age for the Earth as well as, presumably, all of Creation. The remainder of the equation depended on a long-established tradition of biblical chronology.

As early as the second century of the Christian era, Rabbi Jose ben Halafta in his *Seder Olam Rabbah* (Book of the order of the world) listed the dates of biblical events, beginning in Year 0 when God made Adam. Rabbi Jose's "who begat whom when" logic places the births of Seth in *Anno mundi* (the year of the world) 130, Enosh in AM 235, and Kenan in AM 325, and continues through to the births of Methuselah in AM 687, Lamech in AM 874, and Noah in AM 1056. The Flood began in AM 1656, the Exodus from Egypt occurred in AM 2448, the Jewish people entered the land of Canaan after wandering in the wilderness of Sinai for forty years in AM 2488, and the First Temple was destroyed 850 years later, in AM 3338. The modern Jewish calendar continues to follow this logic: the calendar year that began at sundown on September 8, 2010 and continued into September of 2011 was the year AM 5771. Assuming God created Adam six literal days after creating the world, Rabbi Jose's chronology places the Creation at approximately 3760 BCE.

One of the first Christian chronologists, Julius Africanus (c. 170–240 CE), who based his chronology on the Greek version of the Bible, placed the birth of Seth in *Anno Adam* 230, the Flood in AA 2262, the Exodus in AA 3707, and the birth of Christ in AA 5500. Assuming Christ was born in the time frame of approximately 6 to 2 BCE, the chronology of Africanus would place the date of Creation at approximately 5504 BCE.

The most famous early Christian chronology is the *Chronicle* of Eusebius, Bishop of Caesarea in Palestine (d. 339 AD), which we know from Jerome's Latin translation. Eusebius, like Africanus, places the Flood in AA 2262 but dates the Exodus to AA 3689 and the birth of Christ to AA 5199.

But let's get back to our seventeenth-century chronologer James Ussher, who was incidentally the Archbishop of Armagh in Ireland and Vice-Chancellor of Trinity College, Dublin. Ussher published his 1,000+-page *Annals veteris testamenti, a prima Mundi origine deducti* (Annals of the Old Testament, deduced from the first beginning of time) in 1650–54. His datings agreed more closely with those of Rabbi Jose than with either Africanus's or Eusebius's, placing the Flood in AM 1656, the Exodus in AM 2513, and the destruction of the first Temple in AM 3416. In Ussher's chronology, Christ was born in AM 4000. Since he accepted that the year of Christ's birth was 4 BCE, the year for the Creation became 4004 BCE. Ussher was neither the first nor the last scholar to use this biblically-based method to determine the age of the "Heaven and Earth," but he is certainly the best known, most scorned, and least accurately quoted of these scholars. Seen with twenty-first-century hindsight, his methods indeed have little scientific merit. Nonetheless his calculations, in the words of Stephen Jay Gould (1991), "represented the best of scholarship in his time. He was part of a substantial research tradition, a large community of intellectuals working toward a common goal under an accepted methodology . . . ." This form of scholarship used all available historical records to place biblical events into a chronology that included extra-biblical events as well (e.g., the fall of Troy, the founding of Rome, the reigns of emperors and pharaohs, the historical dates of total eclipses of the Sun and Moon). Following the tradition of Jewish chronographers, Ussher placed the beginning of the year in the fall and the beginning of the day at sunset; following Christian tradition, he identified Sunday as the first full day of the first week; and following the most accurate astronomical charts of his day, the Rudolphine Tables of Johannes Kepler, he identified the first Sunday after the autumnal equinox in 4004 BCE as October 23. Therefore, according to Ussher's calculations, the first act of Creation occurred at sunset on the evening of Saturday, October 22, 4004 BCE. Ussher earned lasting fame when the 1703 edition of the King James Version of the Bible appeared with a marginal note identifying 4004 BCE as the date of Creation. That marginal note continued to appear in print in the King James Version well into the twentieth century.

The famous claim that Man was created at nine o'clock in the morning was made by the Reverend John Lightfoot, though it is usually attributed incorrectly to Bishop Ussher. Lightfoot, a contemporary of Ussher, was a distinguished biblical scholar who also became a university vice-chancellor, of Cambridge. In 1642, he published a twenty-page book titled A Few and New Observations upon the Book of Genesis. According to Lightfoot, "Heaven and earth, center and circumference, were created together in the same instant; and clouds full of water . . .were created in the same instant with them. . . . Twelve hours did the heavens thus move in darkness; and then God commanded, and there appeared, light to this upper horizon." That is, for the first twelve hours after Creation, the world was shrouded by darkness. Then, God brought forth the light of day. Lightfoot continues, "Man was created by the Trinity about the third hour of the day, or nine of the clock in the morning." For Lightfoot, the moment of Creation was 6 PM on the evening of the autumnal equinox in 3929 BCE; after twelve hours of darkness and three more hours of light, Man was created at 9 AM the next morning.

Lightfoot's and Ussher's chronologies merged in 1896 in a popular and influential text, A History of the Warfare of Science with Theology in Christendom, written by Andrew Dickson White, the cofounder and first president of Cornell University. In his text, White wrote, incorrectly, that Ussher calculated that man was created at nine o'clock in the morning on October 23, 4004 BCE.

Enter the Astronomers

In the early seventeenth century, Johannes Kepler, whose bona fides as the inventor of mathematical astrophysics via Kepler's laws are unquestioned, proposed an astrophysical model for determining the age of the universe. According to Kepler, at the moment of Creation the Sun would have been at solar apogee (the moment during the Earth's orbit when the Earth and Sun are furthest apart) and simultaneously at the head of the constellation Aries. Kepler understood that the direction of the solar apogee—that is the direction with respect to the background stars in the constellations of the zodiac in which an Earth observer sees the Sun when it reaches its apogee position—changes from year to year, though it does so very slowly. Using then-current values for the rate of movement of the solar apogee and the then-current position of the Sun, Kepler calculated backwards to the time when the solar apogee would have been at the head of the constellation Aries. His answer: God created the world in 3993 BCE at the summer solstice. Kepler's contemporary, the Danish astronomer Christian Longomontanus, used the same logic to calculate the year of Creation as 3964 BCE.

Even Isaac Newton, who coinvented calculus and discovered the first quantitatively successful law of gravity, had his say on the matter. In his *Chronology of Ancient Kingdoms: Amended*, published posthumously in 1728, Newton combined information from scripture, from Herodotus' *Histories*, and from astronomy—he used his calculations of the precession rate of the equinoxes to determine that the Argonaut's expedition took place in 936 BCE—to affix Creation at 3998 BCE.

It would appear, then, that by the early eighteenth century, a scholarly consensus had developed: Creation took place around 4000 BCE, give or take a few decades. Lightfoot (3929 BCE), Longomontanus (3964 BCE), Kepler (3993 BCE), Newton (3998 BCE), and Ussher (4004

TABLE 2.1.

Dates for the Creation of the Universe Based on Biblical Chronology

Chronologist	Date for Creation of the Universe
Rabbi Jose ben Halafta	3760 BCE
Reverend John Lightfoot	3929 BCE
Christian Longomontanus	3964 BCE
Johannes Kepler	3993 BCE
Isaac Newton	3998 BCE
Bishop James Ussher	4004 BCE
Eusebius, Bishop of Caesarea	5203 BCE
Julius Africanus	5504 BCE

BCE) used different scholarly methods and disagreed on some of the details and the exact year, but many theologians, astronomers, and physicists agreed that the Earth and the rest of Creation along with it were only two centuries shy of 6,000 years old. A convergence of opinion is not equivalent to correctness, however, and there were many other scholars, using evidence from other areas of learning, who disagreed with their assertion.

$\cdots\cdot$ CHAPTER 3 $\cdot\cdots$

Moon Rocks and Meteorites

But if the succession of worlds is established in the system of nature,
it is in vain to look for anything higher in the origin of the earth.
The result, therefore, of our present enquiry is, that we find no vestige
of a beginning—no prospect of an end.

—James Hutton, *Theory of the Earth* (1788)

During the seventeenth century, just when an apparent consensus was developing on the age of the Earth, the entire method of biblical chronology came into question. Since chronologists using the three different textual traditions of the Bible (the Hebrew, the Greek, and the Samaritan) obtained time spans since Adam that differed by nearly 2,000 years, the issue of which tradition was most accurate became important. Other scholars asked unanswerable questions that further called into question the method: Was Adam the first man or just the first biblical man? Was the Bible accurate in recording that Methuselah lived for 969 years? Scientists, including the astronomer Edmond Halley, asserted that Scripture did not reveal how long the Earth may have existed prior to the Creation described in *Genesis*. After all, scholars had been arguing for 2,000 years as to whether the days described in *Genesis* were figurative days or literal twenty-four hour days. Perhaps Scripture could reveal the temporal history of humankind on Earth, but it might not provide evidence about the history of the Earth itself, let alone of the entire universe. Furthermore, in a post-Aristotelian world, scientists and philosophers began to think about physical processes through which the world might have been created. Some of these processes, as posited, for example, in the vortex theory of René Descartes (presented in his posthumous publication of 1664, *Le Monde, ou traité de la lumière* [The world, or treatise on light]) and the nebular theory of Immanuel Kant (1755, *Allgemeine Naturgeschichte und Theorie des Himmels* [Universal natural history and theory of the heavens]) and Pierre Simon Laplace (1796, *Exposition du système du monde* [The system of the world]), might have required tens of thousands of years or more.

Fossils

By 1800, Georges Cuvier had invented the science of paleontology and had identified twenty-three species of extinct animals in the fossil record. Fish fossils were found on mountain tops. Fossil mammoths were found in Siberia. Could all of this have happened in less than 6,000 years? Biblical literalists explained the existence of fossils by the theory of catastrophism, which identifies supernatural upheavals like Noah's flood as the causes of cataclysmic changes to life (and rocks) on Earth. This explanation precludes any scientific estimate of the age of Earth and demands that we use biblical chronology to determine its age.

Some members of the emergent scientific community were convinced that "uniformitarianism" could explain the existence of fossils. This theory, first put forth by Scottish geologist James Hutton in 1795 and championed and made popular in 1830 by another Scottish geologist, Charles Lyell, held that geological and biological changes of the past were the results of the same processes that are at work today. This explanation would permit scientists to calculate the age of the Earth based on what they knew of physical, chemical, geological, and biological processes that they could directly observe. Despite finding "no vestige of a beginning," Hutton's geological theory posited an ancient Earth—for Hutton, time could no longer be measured in thousands or even millions of years—but not an eternal Earth. But neither Hutton nor Lyell nor any other geologist in the nineteenth century could venture to suggest a more specific span of years.

Radioactivity

Beginning in the late eighteenth century, scientists began to use concepts from physics to estimate the age of the Earth. For example, in *Les époques de la nature* (1778), Georges-Louis Leclerc, a French naturalist, calculated that if the Earth began as a molten sphere of iron, slowly cooled until it was solid, and then further cooled until it reached its present-day surface temperature, the age of the Earth would be at least 75,000 years and perhaps as great as 168,000 years.

The scientific breakthrough that provides our modern method for determining the age of the Earth came from the discovery of radioactivity by French physicist Antoine Henri Becquerel in 1896. Subsequently, the extensive studies of this phenomenon, including the discovery of the radioactive element radium in the late 1890s by two more French physicists, Marie and Pierre Curie, turned radioactivity into a tool that could function as a geologic clock, a use first suggested by the English physicist Ernest Rutherford in 1905.

Radioactivity is a term that covers several different processes by which atoms fall apart or rearrange their internal structures so as to change from one element to another. To understand radioactivity, we first need to know something about the internal structures of atoms. The universe consists of many elements, all of which are made up of some combination of positively-charged protons, charge-neutral neutrons, and negatively- charged electrons. The number of protons in the nucleus of an atom determines the element. For example, all carbon atoms have six protons in their nuclei. Since protons are positively-charged, a neutral carbon atom must also have six negatively-charged electrons in a cloud that surrounds the nucleus.

Like charges repel each other through the electromagnetic force. Consequently, protons inside a nucleus prefer to be far apart. Protons, however, also attract each other through the strong nuclear force. At the distances between protons in the nucleus, the repulsive electromagnetic force between each pair of protons is stronger than the attractive strong nuclear force between them. As a result, a nucleus made only of protons would be unstable and would fall apart. The nucleus, however, can also contain neutrons that, like the protons, attract each other and attract protons through the strong nuclear force. Since the presence of one or more neutrons increases the total strength of the attractive strong nuclear force within the nucleus without adding additional repulsive positive charges, neutrons moderate the repulsive forces between the protons. Given a sufficient number of neutrons (but not too many), a nucleus with two or more protons can be stable.

The nucleus of a carbon atom cannot hold itself together for more than a few minutes unless the nucleus contains at least six neutrons. Consequently, ^{11}C ("carbon eleven," containing six protons and five neutrons) is unstable because the repulsive force of the six protons on each other is stronger than the attractive force of all eleven of the protons and neutrons for each other. But ^{12}C, ^{13}C and ^{14}C, with six, seven and eight neutrons, respectively, and which are called carbon isotopes (isotopes are atoms with identical numbers of protons but different numbers of neutrons), do exist in nature. If we try to make a ^{15}C atom (with 9 neutrons), it falls apart almost immediately; hence, only three isotopes of carbon exist naturally.

^{12}C and ^{13}C are stable, essentially forever. Carbon fourteen, however, is unstable. Eight neutrons are just too much of a good thing in a nucleus with only six protons. Being unstable, the ^{14}C nucleus eventually undergoes a change; in this case one neutron decays, or falls apart, into a proton plus an electron and an antineutrino in a process triggered by a third force in nature, the weak nuclear force. After the neutron falls apart, the lightweight electron and the antineutrino are expelled from the nucleus while the heavier proton remains, yielding a nucleus with seven protons and seven neutrons. An atom with seven protons is the element nitrogen, and a nitrogen nucleus with seven neutrons and seven protons is "nitrogen-fourteen" (^{14}N). The electron from this decay event (called a *beta* particle) flies out of the nucleus at very high speed and thus carries a tremendous amount of kinetic energy (energy contained in the motion of a particle). Eventually, when this high velocity electron collides with another particle, the enormous amount of kinetic energy it carries (the energy that was generated by the decay process) will be converted into heat energy. This process, in which a neutron falls apart into a proton and an electron, is called *beta decay* and is one of several known radioactive decay processes.

A second kind of radioactive decay process, called *alpha decay*, occurs when a nucleus fissions into two nuclei. One of these will be a small *alpha* particle (a nucleus containing exactly two protons and two neutrons; i.e., a helium nucleus); the other will contain all the other protons and neutrons that were in the original nucleus. The escaping alpha particle will collide with another particle and, as a result of the collision, deposit its kinetic energy as heat. Thus, we can think of both beta and alpha decays as sources of heat.

Radioactive Dating

Radioactivity is a physical process that simultaneously involves incredible precision and complete randomness. On the one hand, our ability to predict the fraction of a sample of radioactive material that will undergo radioactive decay during a specified time interval is extremely accurate. On the other hand, we lack the ability to determine exactly which atoms in our sample will decay.

The steps that allow us to determine the age of an object that contains radioactive materials are as follows. The length of time that will pass before half of our sample of radioactive material will decay is called the radioactive *half-life*. If we have four million radioactive radon atoms (the *parent* species), and if the half-life for this isotope of radon is 3.825 days, then after 3.825 days, half (or two million) of these atoms will have undergone radioactive decay into polonium (the *daughter* species); the other two million radioactive radon atoms will remain in their original condition. Before the clock started counting down the first 3.825 day interval and even during that interval, there would have been no way for us or anyone else to have identified which two million of the four million atoms would decay and which would be stable for that particular time interval. Our inability to identify which 50 percent of the radon atoms will decay is not due to technological limitations but to the very nature of the physics that governs the radioactive decay process. Nevertheless, we could have predicted with high accuracy that about two million of the four million atoms would decay. During the second 3.825 day time period that commences immediately after the first half-life ends, half of the two million atoms that remain in the form of radon after the first half-life will decay to polonium. Thus, after 7.650 days (two half-lives), 75 percent of the original radon atoms (three million) will have decayed to polonium while 25 percent (one million) remain as radon. After a third half-life (a total of 11.475 days), another half-million radon atoms will have decayed to polonium, so that seven-eighths (87.5 percent) of the original population of radon atoms would now be polonium. By determining the elemental ratio of daughter species to parent species, we can determine the length of time that has passed since the population of radon atoms began to decay.

The Radiometric Age of the Earth

When lava flows out of a volcano or magma flows into cracks in pre-existing rocks, the rock that forms from the molten material will have its radiometric clock set to zero. The moment at which the magma solidifies starts the clock. The reason for this is easiest to understand if we consider the example of a radiometric clock such as the decay of potassium forty (^{40}K) to argon forty (^{40}Ar). Since argon is an inert gas, all the argon atoms will bubble out of the magma and escape into the atmosphere. As a result, the solid rock that forms will have some ^{40}K atoms bound into mineral structures within the rock but will have zero atoms of ^{40}Ar. When the potassium decays to argon, the argon atoms are no longer bound into the crystal lattice of the rock. The argon atoms, however, are trapped in the rock like balloons in cages, provided the rock is dense enough and remains relatively undisturbed (e.g., is never heated too much or cracked open). Over time, the ratio of ^{40}Ar to ^{40}K increases from zero (no ^{40}Ar atoms) to one (half of the original potassium atoms are now in the form of ^{40}Ar, half remain as ^{40}K after one half-life) and so on to an ever larger number. By measuring the ratio of ^{40}Ar to ^{40}K atoms in the rock, a geophysicist can measure the time that has passed since this particular rock solidified from a molten state.

By the 1920s, rocks had been found that had radiometric ages of more than 1 billion years, and even a few samples with ages of nearly 2 billion years. In 1921, Henry Norris Russell, the dean of American astronomers and Professor of Astronomy at Princeton University, used the existing information about ages of rocks and the elemental proportions of uranium and lead in crustal rocks "to assign a maximum age for the existing crust of the Earth." He deduced 4 billion years "as a rough approximation to the age of the Earth's crust" and concluded that "the age of the crust is probably between two and eight thousand million years." Twenty years later, theoretical physicist George Gamow reported, in his book *Biography of the Earth*, that the oldest known rocks on the planet were from Karelia (then a Finnish territory; now divided between Finland and Russia), and had an age of 1.85 billion years.

By the beginning of the twenty-first century, rocks older than 3.6 billion years had been identified on every continent on Earth, including the Novopavlovsk Complex in Ukraine (3.64 billion years old), the Morton Gneiss formation in the Minnesota River Valley (3.68 billion years old), the Sand River Gneisses in Zimbabwe (3.73 billion years old), the Narryer Gneiss Complex in Western Australia (3.73 billion years old), the Isua Supracrustal formation in West Greenland (3.75 billion years old), the Imataca Gneiss Complex in Venezuela (3.77 billion years old), the Anshan Complex in northeast China (3.81 billion years old), and the Napier Complex on Mount Stones, Antarctica (3.93 billion years old). The oldest intact rocks yet identified on Earth come from the Acasta Gneiss Complex near Great Slave Lake in northwestern Canada, the oldest of which are aged just over 4 billion years (4.031 ± 0.003 billion years).

While the Acasta rocks are the oldest intact rocks formed on Earth, they are not the oldest rock fragments. As rocks weather and erode, mineral grains known as zircons (zirconium silicate crystals) resist destruction. The mineral grains that surround them may be destroyed by weathering, but the zircons are nearly indestructible. They survive and preserve their precious isotopic and elemental abundances and become incorporated into sedimentary deposits that harden into new rocks. In 2001, zircon grains with ages of 4.3 and 4.4 billion years old were identified in rocks from the Jack Hills rock formation in Western Australia. The Earth therefore must be at least 4.4 billion years old.

Figure 3.1. Zircon grains from Jack Hills, Western Australia. Each grain is a few tenths of a millimeter in length. Image courtesy of Aaron Cavosie, University of Puerto Rico, Mayagüez.

Evidence from the Moon and Meteorites

Zircon grains are the oldest known rock fragments found on Earth that are derived from the weathering of terrestrial rock formations, but even they are not the oldest objects on Earth. That distinction is shared by two categories of objects that are extraterrestrial in origin: lunar rock samples brought to Earth by the Apollo astronauts and meteorites that have fallen to Earth.

Planetary scientists now hypothesize that the Moon formed when a Mars-sized object smashed into the young Earth. The collision splashed out from the outer part of the Earth a small planet's worth of debris that went into orbit around the Earth. Much of that debris quickly accreted in orbit to form the Moon. No other hypothesis offered for the formation of the Moon is able to explain why the Moon has virtually no iron (most of the iron in the Earth had settled to the core before the giant impact occurred), why the ratios of oxygen isotopes in Moon rocks are identical with those ratios in crustal rocks on Earth, and why the Moon is so depleted of volatile substances like water.

The Moon has no atmosphere, no oceans, no life, and no weather. Consequently, the only erosion that occurs is the result of the very slow sandblasting of the lunar surface by cosmic ray particles. In places on the Moon that have not suffered enormous asteroid impacts or been covered in lava flows, rocks have been preserved unchanged from the time they were first formed. Several Apollo missions specifically targeted such locations so that the astronauts might collect samples of rocks that were likely to be extremely old. Quite a few lunar rock samples from the lunar highlands have radiometric ages of at least 4.4 billion years. Several Moon rocks brought back by the Apollo 15, 16, and 17 astronauts have radiometric ages of 4.5 billion years. Since the Moon must be older than the oldest rocks on the lunar surface, the Moon is at least 4.4 and almost certainly at least 4.5 billion years old. Since the Earth is older than the Moon, the Earth must be at least 4.4

Figure 3.2. Ancient lunar crustal rocks collected on Apollo missions. These rocks are about 4.5 billion years old, indicating that parts of the Moon's crust solidified soon after the Moon was formed. Photo courtesy of NASA Johnson Space Center.

Figure 3.3. The Allende meteorite. The white to light gray, irregularly shaped speckles in the Allende meteorite are calcium-aluminum-rich inclusions (CAIs) Photo courtesy of Laurence Garvie/ASU; courtesy of Center for Meteorite Studies, ASU.

billion years old and probably more than 4.5 billion years old, which is a few tens of millions of years and perhaps 100 million years older than the oldest terrestrial zircon grains known today.

Virtually all meteorites have radiometric ages that fall between 4.4 and 4.6 billion years. The very oldest materials in the oldest meteorites are tiny mineral structures known as calcium-aluminum-rich inclusions (CAIs). Like zircon grains, CAIs have remained intact since their formation from material in the gaseous and dusty disk surrounding the young Sun. Over time, they collided with other small objects. Through these sticky collisions, the CAIs were incorporated into bigger objects that grew from dust-grains into pebbles, rocks, and boulders. These larger objects continued to orbit the Sun. Those that remain we call asteroids.

Every year, some asteroids collide or exert enough gravitational attraction on each other to change each others' orbits so that a few of them, or fragments of a few of them, end up in orbits that intersect the orbit of the Earth. Some of these so-called "near Earth asteroids" eventually collide with the Earth. When the Earth and an asteroid collide, some parts of the asteroid burn up in the atmosphere while other parts may survive passage through the atmosphere and land on the ground without being melted or vaporized. Such objects are known as meteorites. The oldest CAIs, which are found only in certain kinds of fragile meteorites known as carbonaceous chondrites, have been found in the meteorite Efremovka, which fell in Kazakhstan in 1962. The ages of Efremovka's CAIs are 4.567 billion years (four billion, five-hundred-sixty-seven million years). Therefore, if the planets orbiting the Sun formed after the first meteorites formed in orbit around the Sun, which is likely, but before the Moon formed, which also is likely, the Earth must be older than 4.5 billion years and younger than 4.567 billion years. Since the universe, of course, must be older than the Moon, the Earth, and the oldest meteorites, these ages put a lower limit on the age of the universe.

···CHAPTER 4···
Defying Gravity

It seems, therefore, on the whole most probable that the sun has not illuminated the earth for 100,000,000 years, and almost certain that he has not done so for 500,000,000 years. As for the future, we may say, with equal certainty, that inhabitants of the earth can not continue to enjoy the light and heat essential to their life for many million years longer unless sources now unknown to us are prepared in the great storehouse of creation.

—Sir William Thomson (Lord Kelvin), in "On the Age of the Sun's Heat," *Macmillan's Magazine* (1862)

Over the course of the twentieth century, geologists and geochemists slowly teased the secrets of the ages from Earth rocks, lunar samples, and meteorites. But we should keep in mind that even our very best estimate for the age of the Earth is only a lower limit. Perhaps no rocks survived the first 1 or 5 or 30 billion years of Earth history. Perhaps the Moon formed long after the Earth. Perhaps no ancient Moon rocks have survived, or perhaps the Apollo astronauts did not visit the most ancient lunar rock formations and thus did not bring back the oldest Moon rocks. Perhaps older meteorites exist but are too fragile to survive the fiery descent to Earth's surface. Perhaps the very oldest asteroids orbit the Sun in a part of the solar system from which few or no meteorites are delivered to Earth.

If any of these "perhaps . . . " scenarios is correct, then the radiometric ages obtained from Earth and Moon rocks and meteorites are only coincidentally coeval. And if that is the case, then the radiometric age of about 4.5 billion years we have obtained for the oldest objects yet studied might not tell us very much about the age of the Earth or the age of the Sun, let alone the age of the universe. On the other hand, the Earth, the Moon, and the meteorites may all have formed at about the same time, about 4.5 billion years ago. But what is the relationship between the age of the Sun and the age of the objects in orbit around it? If the Sun formed first and then, at a much later time, captured fully formed planets from interstellar space, there would be no relationship between the ages of the Sun and its planets. Alternatively, the Sun may have formed coevally with its planetary system. If we had a way to determine the age of the Sun, we would know the relationship between the measured ages of Earth rocks, lunar rocks, and meteorites and of the solar system as a whole. Determining the age of the Sun seems like a sensible next step as we work toward measuring the age of the universe.

In the eighteenth century, both Kant and Laplace suggested that the Sun and the planets and smaller objects that orbit it could have formed from a swirling interstellar cloud. Two centuries later, astronomers confirmed this hypothesis by identifying many such interstellar clouds that exist today and by studying the star formation process that takes place within these clouds. In general, interstellar clouds are in a precarious balance between expansion and collapse. Their internal heat generates expansive pressure while gravity from the particles of matter in the cloud works to pull the particles closer together. When an interstellar cloud cools off, its thermal pressure weakens and the cloud loses the ability to resist the compressive force of gravity. The balance tips in favor of gravity and the cloud collapses in on itself. The rotational motion of the collapsing cloud, however, prevents all of the material in the cloud from falling all the way to the center; instead, while the cloud shrinks it also flattens, forming a disk of gas and dust that revolves around the newborn star at its center. The planets that orbit our Sun formed in just such a disk.

We now understand the physics of the process of star formation well enough to know that it involves a set of associated events. The physical processes involved in the gravitational collapse of an interstellar cloud are such that the smallest particles in the disk could be almost as old as the central star; the moons and planets would form from the disk

a.

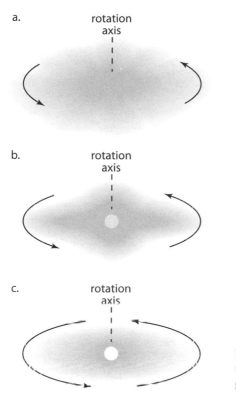

Figure 4.1. Illustration of the theory that stars form from rotating clouds of gas in space.

particles within a few tens of millions of years after the star started to form. All the astrophysical evidence supports the assertion that the Sun is older than the Earth, but only by a few million to a few tens of millions of years. The oldest meteorites could be among the very first solid objects formed in the solar system and might be almost as old as the Sun itself.

The Sun Must Have an Energy Source

The Sun is a large spherical object, made mostly of hydrogen and helium gas, that emits light. That light heats and illumines the Earth. We know the distance of the Earth from the Sun and we know the physical size of the Earth. From these two numbers, we can calculate the

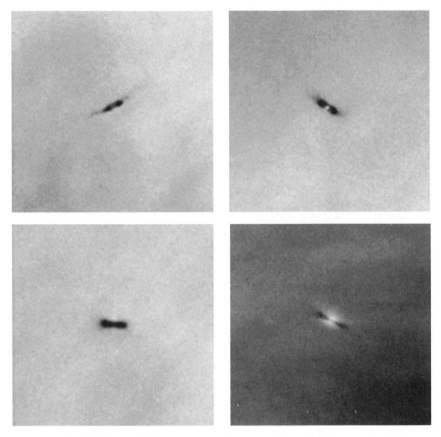

Figure 4.2. Hubble Space Telescope images showing flattened disks around newborn stars (known as *proplyds*) in the Orion Nebula. Image courtesy of NASA, ESA and L. Ricci (ESO).

fraction of emitted sunlight that is intercepted by the Earth. And from all of this information, we can calculate the total amount of energy released by the Sun every second. Simply by measuring the distances and luminosities of other stars, we can perform similar calculations for any star in the sky.

If stars radiate heat, their surface temperatures should decrease unless the heat is replenished from within. Since the interiors of stars will be hotter than their surfaces, stars will compensate for any heat lost from their surfaces through the transfer of heat from their centers to their surfaces. Unless the cores of stars have heat sources, they must cool off. If the cores cool off, the entire insides of stars should eventually cool off and the stars should contract. If contraction occurs, it should be measurable on the time scale of a human lifetime, and substantial changes should have been evident over recorded history.

The surface temperature of the Sun is not decreasing, however; and historical evidence gives no indication that the Sun has cooled over human history or even over geological history. We also see no evidence to suggest that the surface temperatures of any other stars are decreasing with time. If neither the Sun nor other stars are cooling, they therefore must be capable of generating heat from internal sources to replenish the heat radiated to space.

In the mid-nineteenth century, the German physicist Hermann von Helmholtz conjectured that if the Sun derived its heat from the oxidation of combustible materials like wood or coal, it could burn for only 1,000 years. If, however, the Sun were slowly contracting, with its outer parts settling inwards toward its core, it would generate the energy required to shine for a much longer period by converting gravitational potential energy into heat. This is the same process that heats a metal spike when you drop the head of a sledgehammer repeatedly onto the spike. Once the sledgehammer is lifted off the ground, it contains potential energy. When the head of the hammer is released, that potential energy is converted into the energy of motion (kinetic energy), which causes the hammer to drop toward the spike. When the hammer hits the spike, that kinetic energy is transferred to the spike. Some of the kinetic energy drives the spike into the ground, while some simply makes the iron atoms in the spike vibrate faster. The energy associated with the individual motions of the atoms in the spike produces what we call heat. The faster the motions of the atoms the higher the temperature of the spike. When the Sun contracts, atoms more distant from the center of the Sun fall inwards and collide with atoms slightly less distant from the center, thereby converting gravitational potential energy to kinetic energy; the conversion of potential to kinetic energy heats the outer layers of the Sun. According to Helmholtz's calculations, this process could generate enough heat to allow the Sun to shine for 20 to 40 million years. Lord Kelvin, the great English physicist and a contemporary of Helmholtz, repeated

Helmholtz's calculations and found that the Sun could be as much as 500 million years old. Presumably the Earth could be equally old. This age might be sufficient to permit rocks to form according to the natural processes outlined in Hutton and Lyell's theory of uniformitarianism. These nineteenth-century attempts at estimating the maximum length of time for which the Sun might continue to shine provided plausibility for the hypothesis that the Sun, and by inference the Earth, might be older than 6,000 years, but they did not provide an actual age for the Sun.

If the mechanism of gravitational contraction is indeed generating the Sun's energy, as Helmholtz and Kelvin proposed, that process has a testable, observable consequence: according to Lord Kelvin's calculations, the Sun should contract in diameter by about 70 meters per year. While measuring such a small change in the solar diameter was beyond the ability of nineteenth-century astronomers, such measurements are well within our twenty-first century capabilities, and we now know that the Sun's diameter is not changing. Gravitational contraction does not power the Sun.

$E = mc^2$

If the Sun is not changing measurably in luminosity or temperature at either its surface or its core, if the Sun is not contracting, and if the Sun is billions of years old (as it must be to match the age of the Earth), then it must have an enormously powerful internal source of energy that is able to replenish the heat lost from the core to the surface and from the surface into space. No nineteenth-century theory could identify this energy source.

English astrophysicist Arthur Eddington, in 1926, proposed a new method for energy generation in stars that was based on Albert Einstein's theory of special relativity. One tenet of special relativity is that mass (m) is equivalent to energy (E) and that the amount of energy contained by a piece of mass is found by multiplying the mass by the square of the speed of light (c^2). That is, $E = mc^2$. Effectively, $E = mc^2$ expresses two ideas: that mass is simply one way in which the universe stores energy; and that energy can be converted from one form to another if the physical conditions (temperature, density, pressure) are right. Eddington suggested that four hydrogen nuclei (four individual protons) could be combined, or fused together, to make one helium nucleus in a process called *nuclear fusion*. Since the mass of one helium nucleus is slightly less than the sum of the masses of four protons, Eddington suggested that the "lost" mass had been converted to energy, and that it was this energy that powered the stars.

In 1929, Henry Norris Russell was able to calculate the relative amounts of the elements in the atmosphere of the Sun and concluded that more than 90% of the Sun, by volume, and about 45% by mass, must be composed of hydrogen. Stars therefore had nearly inexhaustible supplies of hydrogen and so, according to the process sketched out by Eddington, could power themselves for billions of years. Aided by the discovery of the neutron in 1932 by James Chadwick and the development of the theory of quantum mechanics in the 1920s and 30s, Hans Bethe deduced the sequence of nuclear reactions that takes place in the cores of stars.

The Energy Available through Nuclear Fusion

In a sequence of reactions called the proton-proton chain, four protons (^{1}H) combine to form a single helium nucleus (^{4}He) made of two protons and two neutrons; however, this does not occur via the highly improbable simultaneous collision of four particles. Instead, the proton-proton chain involves several intermediate steps and six, not just four, protons. First, two protons collide. After the collision, one of the protons is converted into a neutron through the emission of two particles, a positron (the anti-particle partner of an electron, having the mass of an electron but a positive charge) and a neutrino (a very low-mass particle with no electric charge). The resulting particle contains both a single proton, which means that it is still a hydrogen nucleus, and a neutron, which makes the nucleus heavier than a normal hydrogen atom. This heavy hydrogen atom is known as *deuterium*, which is denoted as either ^{2}H or D.

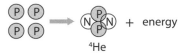

Figure 4.3. In the proton-proton chain, four protons combine to form a helium nucleus. In the process, some mass is converted into energy and a small amount of mass is converted into particles called neutrinos. This nuclear fusion reaction generates the energy that powers the Sun.

The positron will quickly find its anti-particle, an electron; and in that collision, they will annihilate one another, turning all of their combined mass into energy in the form of a high-energy gamma ray photon. The gamma ray doesn't travel far before it is absorbed by another particle, adding to that particle's energy and thereby making it move faster. Since the average speed of the particles in a gas determines its temperature, when this first step in the proton-proton chain has been repeated many times it has the effect of heating the gas at the center of the star. The neutrino has properties such that it only rarely collides with other particles (it is known as a weakly-interacting particle), so almost all neutrinos produced in this reaction fly right out of the Sun.

In the next step in the proton-proton chain, the deuterium nucleus collides with another proton to form a helium nucleus, though this is a lightweight helium nucleus with two protons but only one neutron (^{3}He; called "helium-three"). This reaction also generates a gamma ray, which will be absorbed by a nearby particle, contributing excess kinetic energy to that particle and heating the surrounding gas.

These first two reactions must each happen twice so that two ^{3}He nuclei are created. Finally, these two ^{3}He nuclei collide, forming a ^{4}He nucleus and knocking loose two protons. The combined mass of four protons is 6.690×10^{-24} g, while the mass of one ^{4}He nucleus is 6.643×10^{-24} g. The fractional difference in mass between the input and output particles, equal to 0.7 percent of the starting mass, is the amount of mass converted to energy in this process. If the entire mass of the Sun were available (which it is not) for this mass-to-energy conversion process, the proton-proton cycle could power the Sun for 100 billion years.

Critical Requirements for the Proton-Proton Chain

The collisions that power the proton-proton chain involve positively-charged nuclei colliding with other positively-charged nuclei. Positively-charged particles repel, however, so two protons (for example) are unlikely to collide except under the most extreme conditions. In fact, if two protons were propelled toward each other at low speeds, the electromagnetic repulsion exerted by each on the other would prevent the collision from happening, just as two automobile drivers driving toward each other on a single-lane country road at low speeds would likely see each other in time to avoid the collision either by slamming on their brakes or veering out of each other's way.

Let's follow the potential car crash analogy further. Under what circumstances might the two drivers be unable to avoid a collision, either with each other or with another innocent-bystander car? We can identify two preconditions that would certainly raise the likelihood of such a collision: high speeds and a high density of cars. High speeds make it less likely that a driver will have enough time to react after discovering another car in his path; high density—meaning that the parking lanes on both sides of the narrow road are packed with other cars—virtually ensures that any effort to avoid a collision with a car in the driver's lane will almost certainly cause a collision with another, nearby vehicle. If both the high speed and high density conditions are met, a collision between the two cars is inevitable.

In order for two protons to collide, they must get close enough to touch; that is, they must come closer than one nuclear diameter (10^{-13} cm). The minimum temperature required for two protons to overcome their mutual repulsion at this close distance is 10 billion K. This implies that while the surface of the Sun has a temperature of only 6,000 K, the core must have a temperature about one million times hotter if nuclear fusion is the process that powers the Sun; yet astronomers were certain, even in the 1920s, that the core of the Sun could not be much hotter than about 10 million K, which is about 1,000 times cooler than 10 billion K. They reasoned that if the core temperature of the Sun were much hotter than 10 million K, the intense pressure from the super hot gas deep inside the Sun would cause the outer layers of the Sun to expand, increasing its size much above its present diameter.

Clearly, if stars generate energy through nuclear fusion but do so at temperatures of millions of degrees rather than billions of degrees, then our simple picture of the fusion process is so far inadequate. We need to furnish two additional and crucial links. One is the consequence of what is called the *kinetic theory of gases*. In a gas—and the particles at the core of the Sun are indeed in a gaseous state—every particle moves at a different speed. Some particles move slowly in comparison to the average; others much faster. When we ask, what is the temperature of this gas? we are actually asking, what is the average speed of all the particles of which it is composed? In this distribution of velocities, known as the Maxwell-Boltzmann distribution, some particles will be moving twice as fast, some as much as six times as fast as the average. So, for example, if the temperature of the gas is 10 million degrees, a very small fraction of its particles are moving with speeds equivalent to the average velocity of particles in a gas at 60 million degrees. Thus, we do not need the temperature of the gas to be 10 billion K in order for a few particles within it to be moving at speeds as fast as the average for a 10 billion degree gas. But we do need temperatures well above 10 million K. The kinetic theory of gas makes nuclear fusion more possible, but by itself it is not sufficient to explain how fusion can take place in the Sun.

The second missing link is called *quantum tunneling*, an idea discovered by George Gamow and also, independently, by Ronald W. Gurney and Edward Condon in 1928. In the case of two protons, we can think of the electromagnetic repulsion one proton exerts on the other as a sort of energy barrier or hill that the second proton must jump over in order to bump into the first proton. If the second proton is moving fast enough, it is able to climb the energy barrier between the two protons and have enough energy left over to collide with the first proton on the other side. The quantum-tunneling concept says that there is a small

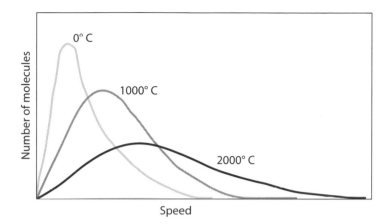

Figure 4.4. The Maxwell-Boltzmann distribution depicts the speed (x value) versus the number of particles in a gas moving at each speed (y value). As the gas temperature increases, the entire distribution broadens and shifts toward higher speeds. In the core of the Sun, only a very few particles with the very highest speeds have enough energy to participate in nuclear fusion reactions.

but real possibility that the second proton can get past—metaphorically, tunnel through—the barrier despite lacking the energy to climb over. The likelihood that two protons will collide at temperatures of only about 10 million K, without quantum tunneling, is effectively zero. But quantum-tunneling calculations indicate that though this event is unlikely, it should happen for any pair of protons once in about 10 billion years. This probability, restated, means that if I have ten billion protons, this event should happen once per year. Since the core of the Sun has an almost unfathomable number of protons (about 10 to the 55th power, 10^{55}), enough of these unlikely collisions happen every second (in fact more than 10^{38} per second) that the Sun is able to power itself through nuclear fusion reactions.

If we start at the core of the Sun and move outwards, both the temperature and density of the gas will decrease toward the surface. Outside of a critical radius, the temperature and density will be too low for any nuclear fusion reactions to take place, even with the help of quantum tunneling. The region inside this critical radius is the core of the Sun; the region outside the core is called the envelope. In the Sun, only the hydrogen located in the core can participate in the proton-

proton chain; the hydrogen in the envelope is inactive in the nuclear fusion process.

How Long Can the Sun Power Itself from Proton-Proton Chain Reactions?

Eventually, the Sun will run out of protons in its core and will no longer be able to fuel the proton-proton chain reactions. This eventuality necessarily will lead to changes in the behavior of the Sun. When energy is no longer generated in the core, the heat radiated from the surface will no longer be fully replenished from inside the star. The entire Sun will begin to cool off and contract. As we will see in Chapter Thirteen, the cooling and contraction of the core of a star will lead to a heating and compression of the core, which in turn will lead to a new set of nuclear reactions that occur at higher temperatures and pressures. Had those new nuclear reactions already begun inside our Sun, they would have produced internal changes in its structure. Those internal structural changes would cause the Sun to increase in size until it became a red giant star. As a red giant, the Sun would be larger, more luminous, and cooler at its surface. But the Sun is not a red giant, yet; therefore, it has not exhausted the supply of protons in its core.

The core of a typical star comprises about 10 percent of its total stellar mass, and about 0.7 percent of that mass can be converted to energy via proton-proton chain reactions. If we calculate the total amount of energy that can be generated by converting 0.7 percent of the mass of the core of the Sun from protons into helium nuclei, and if we divide that number by the luminosity of the Sun, we come up with the length of time during which the Sun can shine as it does today: approximately 10 billion years. We therefore may be confident that the Sun is less than 10 billion years old. But can we pin down its age more accurately?

Every second, the Sun converts a huge number of protons into helium nuclei. These conversions affect the density, temperature, and pressure in every layer of the Sun, from its center all the way out to the surface. Since the rate of nuclear reactions in the core depends on density, temperature, and pressure, these changes have a feedback effect on the nuclear fusion process itself. Slowly and steadily, these changes accumulate and affect the luminosity and temperature of the surface of the Sun, making the Sun a little bit hotter and brighter over a timescale of a few billion years. Given the mass and composition of the Sun, astrophysicists can calculate what the luminosity and surface temperature of the Sun should have been when it was born and how those parameters should evolve with time. From those calculations, we know that the Sun is neither newborn nor nearing the end of its lifetime; in fact, the Sun is about 4.5 billion years old. Were it younger, it would be cooler and less luminous. Were it older, it would be hotter and more luminous.

Our understanding of the astrophysics of the Sun has led to the conclusion that the Sun is about the same age as the oldest meteorites in the solar system. This result is independent of and consistent with our observations that stars and their planetary systems form at about the same time. We can say with confidence that the Sun and all the objects in orbit around it, from the tiniest meteorites to the Moon, Earth, and other planets and moons, all formed very nearly 4.5 billion years ago and that the universe is therefore at least that old.

In order to determine whether the rest of the universe is also 4.5 billion years old or is older, we will need to carry our investigation far beyond the confines of our solar system. Anyone who has gazed into the nighttime sky knows that the dominant objects visible to our unaided eyes are stars. Perhaps they can teach us more about the age of the universe.

• : 2 : •

THE AGES OF THE OLDEST STARS

$\bullet \bullet \bullet$ CHAPTER 5 $\bullet \bullet \bullet$

Stepping Out

It is the greatest and most glorious triumph which practical astronomy has ever witnessed.

—John Herschel, President, Royal Astronomical Society, in his "Address Delivered at the Annual General Meeting of the Royal Astronomical Society, February 12, 1842, on Presenting the Honorary Medal to M. Bessel," *Memoirs of the Royal Astronomical Society* (1842)

To find out if we can learn more about the age of the universe from stars, we first need to know something about the stars themselves. What are they? We know of course that they are sources of light. So it stands to reason that if we come to understand more about the nature of light, we might use that knowledge to expand our understanding of the stars themselves. Among the most fundamental questions we might then ask about the stars (which we will indeed ask in Chapters Five and Six) are: How much light does each star emit? and How bright is that star? Later we will discover that by calculating the amount of light a star emits at different colors, we can measure its temperature (Chapter Seven) and determine its size (Chapter Eight). From the temperatures and brightnesses of stars, Henry Norris Russell and Ejnar Hertzsprung will cobble together the most important diagram in all of astrophysics (Chapter Nine). Then they and other astronomers will learn how to use the motions of stars in binary star systems to measure stellar masses (Chapter Ten) and how to use this diagram and observations of star clusters to measure distances to star clusters (Chapter Eleven). In the early and mid-twentieth century, astronomers will employ spectral measurements to determine the elemental constituents of stars and nuclear physics to determine how stars produce energy. In combination, these tools enable astronomers to figure out how stars are born, live, and die (Chapter Twelve). In turn, understanding the life-times and life cycles of stars, will enable astronomers to determine the ages of white-dwarf stars (Chapter Thirteen) and star clusters (Chapter Fourteen), which will at last lead us to consistent estimates for the age of the universe.

The Apparent Brightnesses of Stars

The brightness that we are able to measure for any object—astrono-mers call this the *apparent brightness*—is not the same as the intrinsic, or absolute, brightness of that object. Apparent brightness depends on two properties of that object: its intrinsic brightness and its distance from us. Astronomers can easily measure the apparent brightness of a

Figure 5.1. The Big Dipper. Some stars appear brighter than others; however, the brightest stars may not be intrinsically brighter than the fainter stars. Instead, they may be intrinsically faint stars that happen to be near the Sun. Image courtesy of Noel Carboni.

star by using a counting device (a photographic plate, a photocell, or a CCD camera) to calculate the number of photons that arrive on Earth every second from a given star. If we were able to measure the distance to that star, we could combine that information with our measurement of the star's apparent brightness and arrive at its intrinsic brightness. With that information in hand, we might be able to learn even more about the star; eventually, we might be able to determine its age. Our next step, therefore, is to figure out how to measure the distances to stars.

The Size of the Earth as a Measuring Stick

In the third century BCE, the Greek geometer and astronomer Eratosthenes crafted the first astronomical measuring stick when he calculated the circumference of the Earth. He achieved this feat by measuring the angle between the Sun and the zenith (the local vertical direction, at any location on Earth) in two cities in Egypt, one located almost directly north of the other, at noon on the day of the summer solstice. One measurement was made in the town of Syene (now Aswan). There, the Sun was directly overhead at noon, so that the direction to the Sun was the same as the direction to the local zenith. His other measurement was made in Alexandria, located directly north of Syene, where the angle between the Sun and the zenith at noon was 7.2° (one fiftieth of the circumference of a circle). Assuming the Earth was spherical, Eratosthenes used this angle, combined with basic geometric principles, to deduce that the distance from Alexandria to Syene must be one fiftieth of the circumference of the Earth. By pacing off the distance between the two cities and multiplying by fifty, he reckoned the circumference (or equivalently, the diameter) of the Earth to an accuracy within a few percent of our modern value. Using other geometric arguments, he then measured the distance to the Moon and Sun in units of Earth diameters and again achieved answers within a few percent of the values obtained by modern measurements. It was now conceivable that the physical distances that separated our Earth from all other objects visible in the heavens might one day be

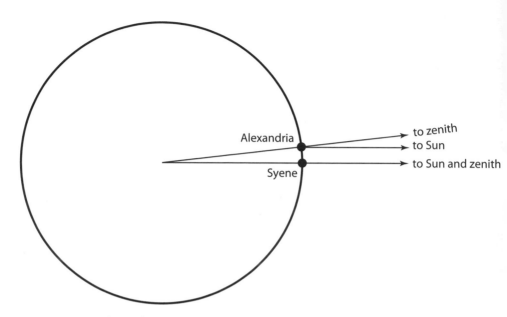

Figure 5.2. Eratosthenes' method for measuring the circumference and diameter of the Earth.

measureable. With Eratosthenes humankind had taken its first step toward surveying the universe.

The Astronomical Unit:
A Bigger Measuring Stick

Almost two millennia later, astronomers took yet another crucial step into the cosmos. They determined the length of the *astronomical unit* (AU).

The astronomical unit is a measure of length related to the size of the Earth's orbit. From ancient times, astronomers had attempted to measure the sizes of the presumed-to-be-circular orbits of the known planets in proportion to the size of the Sun's presumed-to-be circular orbit, with the Earth being, by definition, one AU from the Sun. In the sixteenth century, Nicholas Copernicus applied his new heliocentric theory of the solar system to calculate the relative sizes of the orbits of Mercury, Venus, Earth, Mars, Jupiter, and Saturn. In spite of his assumption that the planets orbited the Sun in circular orbits, he obtained answers that were correct to within 4 percent. Early in the seventeenth century, Johannes Kepler deduced that planets orbit the Sun in elliptical rather than circular orbits (this is known as Kepler's first law). Simply put, ellipses are flattened circles characterized by one short axis (the minor axis) and one long axis (the major axis). Half of the length of the major axis is the semi-major axis. The astronomical unit, now redefined by Kepler, became the semi-major axis of the Earth's elliptical orbit rather than the radius of the earth's circular orbit and one of the two key parameters in his third law, which relates the semi-major axis to the orbital period for any object orbiting the Sun. Although Kepler was able to determine the relative sizes of the other planets' orbits quite accurately from their measured orbital periods, those distances could be stated only in astronomical units; for example, the semi-major axis for the orbit of Jupiter was 5.2 AU, for Saturn 9.5 AU. Kepler, however, could only guess at the actual physical length of the astronomical unit.

In 1672, the French astronomers Jean Richer and Jean-Dominique Cassini made the first reasonably accurate measurement for the physical length of the astronomical unit by measuring Mars' *trigonometric parallax* (a concept that we will tackle in the next section). Their answer: 140 million kilometers. By the end of the nineteenth century, astronomers had refined this measurement to an accuracy of better than 0.1 percent. Finally, by bouncing radar signals off the surfaces of Venus in 1961 and then Mercury in 1962, astronomers arrived at a value of the astronomical unit that was accurate to one part in one hundred fifty million. With the value of the astronomical unit (149,597,870.69 km) now fixed with such precision, we know with equal precision the distance from the Sun of any object with a well-determined orbit around the Sun.

So far, then, our reach into the cosmos extends to all objects orbiting the Sun. But how far exactly is that? Pluto, the once and perhaps future ninth planet, has an orbital semi-major axis of 39.5 AU. Sedna,

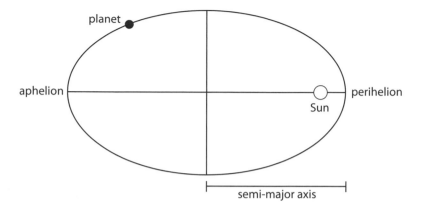

Figure 5.3. Kepler discovered that planets orbit the Sun in ellipses, not circles. The size of an orbit is typically described in terms of the semi-major axis of the ellipse (half the length of the long axis). For the Earth's orbit, this length is one astronomical unit. The closest approach of the planet to the Sun is called the *perihelion point*, the point at which the planet is furthest from the sun is the *aphelion point*.

which most likely is an object scattered out of the Kuiper Belt after a close gravitational encounter with another Kuiper Belt object and which is currently just under 90 AU from the Sun, has an orbital semi-major axis of 536 AU. Because Sedna has a highly elliptical orbit, it will come as close to the Sun as 76 AU in 2075 and will travel out to 975 AU in about 6,000 years. Some comets have orbits that take them much further from the Sun than Sedna, out to distances beyond 10,000 AU from the Sun. These comets require millions of years to complete a single orbit. Yet, the incredible distance to Sedna and to these distant comets pales in comparison to the distance to Proxima Centauri, which is the closest star (other than the Sun) to the Earth. And of course Proxima Centauri does not orbit the Sun, so we cannot measure the distance to Proxima Centauri or to any other star using the methods we used to survey distances within our solar system.

To step further out into the universe and measure distances to stars, astronomers needed a tool more far-sighted than the astronomical unit. Knowledge gained from the astronomical unit did, however, allow them to develop the technique that would bring them closer to their goal. Namely, trigonometric parallax.

Trigonometric Parallax

Hold your arm straight in front of you with your index finger pointed up. Close your right eye and look at your finger with your left eye, making a mental note of the positions of more distant objects that appear nearly directly behind your index finger. Now open your right and close your left eye; again, make a mental note of the positions of more distant objects seen directly behind your index finger. If you did not already know that your finger was nearby and the background objects much more distant, you would think that your finger had moved from one location to another; in fact, your finger never moved but the location (first your left eye, then your right eye) from which you made your measurement did. This apparent change in the position of your finger is called *trigonometric parallax*. With binocular vision, we can notice small angular shifts (called the parallax angle) in the apparent positions on objects. Because closer objects have larger angular shifts than more distant objects, the relative sizes of trigonometric parallax angle measurements enable us to infer relative distances.

Parallax measurements can be made precise and quantitative if we think about these measurements in terms of the sides and angles of a right triangle. A right triangle has one 90-degree angle. When we make a parallax measurement, we are measuring one of the other two angles. Since the sum of the three angles in a triangle is 180 degrees, our one measured angle yields the values of all three angles of our right triangle. Applying laws of geometry developed by the Greek geometer Euclid

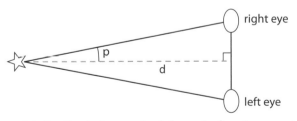

Figure 5.4. Parallax is the perceived change in direction or position of an object when we view that object from two different locations. We do this all the time using our binocular vision (as each of our eyes is at a slightly different viewing location).

2,400 years ago, we can calculate the lengths of all three sides of a tri-angle for which we know the value of any two angles (implicitly, we know all three angles if we know any two) and the length of any one of the sides. Consequently, we can measure the distances to stars if we can succeed in measure the parallax angle to a star (in practice no easy feat) and the length of one side of our parallax triangle.

When we make parallax measurements with our eyes, we are actu-ally making two measurements, one with each eye; and we are making use of two right triangles placed back-to-back. The two triangles are identical in size, mirror images of each other, and they share one side: the distance from the bridge of our nose to the object we are observing. In this case, we also know the length of one of the sides, namely the base of each triangle: the distance from the bridge of our nose to the center of one eye socket, which is about 3.5 centimeters. Given that we know all three angles (the right angle, the measured parallax angle, and 180 degrees minus the summed value of these first two angles) and the length of one side of our triangle, we can calculate the physical distance to the object of interest. Our brains carry out these trigono-metric parallax measurements and calculations constantly, enabling us to judge distances and (for some of us) safely parallel park our cars or catch a baseball.

Our ability to distinguish parallax angles for objects we observe di-minishes as the distances to the objects increase. At some limiting dis-tance, the parallax angle becomes immeasurably small, so minute that we conclude the objects are far away; we can determine neither the actual distances nor which of two such objects is the more distant.

We can make accurate parallax measurements of more distant ob-jects if we make our triangle bigger. Since we cannot change the dis-tance to a star, the only part of the triangle we can change is the length of the base. We can increase the baseline length by making mea-surements with two telescopes located at two widely separated sites rather than with our two eyes. When we make parallax measurements with two widely separated telescopes, our baseline is half of the dis-tance between the two observation points. If we locate our two tele-scopes on different continents, we can separate them by a few thousand kilometers.

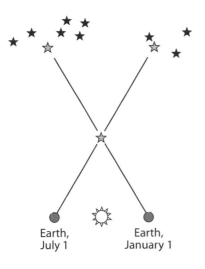

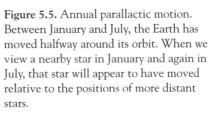

Figure 5.5. Annual parallactic motion. Between January and July, the Earth has moved halfway around its orbit. When we view a nearby star in January and again in July, that star will appear to have moved relative to the positions of more distant stars.

Earth,
July 1

Earth,
January 1

But we can do much better than that. All we need do is take advantage of a very simple property of the Earth: it orbits the Sun. By analogy with the geometry of our faces, imagine that one "eye" is a telescope on or in orbit around Earth at any given moment in time and that the second "eye" is the same telescope located on or in orbit around Earth exactly six months later, when the orbital motion of Earth has carried our telescope to the other side of the Sun; the Sun's location is analogous to the bridge of our nose. The "eye to nose" distance for our new triangle is the distance from the Earth to the Sun, i.e., the astronomical unit. We have now achieved a triangle base of nearly 150 million kilometers, which can be used to measure the parallax angles to stars located at very large distances. Eventually, astronomers hope to increase the baseline even further by placing telescopes in distant parts of the solar system, but for the foreseeable future, our parallax measurements will be limited by the size of Earth's orbit.

The Parsec—An Even Bigger Measuring Stick

When astronomers work out the geometry for parallax measurements, the mathematics can be made very simple and clear if we choose the right units for distances and angles. First, angular units. Recall that the angular distance around the circumference of a circle is 360 degrees. Each degree is composed of 60 minutes of arc (when you look up at the Full Moon, the angular size of the Moon is about 30 minutes of arc). Each minute of arc is composed of 60 seconds of arc. In total, the circumference of a circle is equal to 1,296,000 seconds of arc. If we measure the parallax angle for a star and find that the angle is exactly one second of arc, and if we then do the calculations, we find that the distance to that star is 206,265 AU. Since this is a big, unwieldy number, astronomers find it convenient to define this distance as 1 *parsec*; thus, a parsec is the distance at which a star with have a *parallax* angle of one *second* of arc. With this new unit defined, the distance to a star (in parsecs, abbreviated as "pc") is found as the inverse of the parallax angle (in seconds of arc). For example, if the parallax angle is 0.1 seconds of arc (written as 0.1″), the distance to the star is 10 parsecs. Or, if the parallax angle is 0.01″ the distance to the star is 100 parsecs.

As is evident from these two examples, if the parallax angle is large, the distance to the star is small; if the parallax angle is small, the distance to the star is large; and if the parallax angle is too small to measure, the distance to the star is too large to measure using the technique of parallax. For more distant objects, we define the units *kiloparsec* (1 kpc = 1,000 pc), *megaparsec* (1 Mpc = 1,000 kpc), and *gigaparsec* (1 Gpc = 1,000 Mpc).

Real Parallax Measurements: Not as Easy as It Sounds

For 2,000 years, ever since the ancient Greeks first began measuring the motions and locations of objects in the heavens, astronomers made parallax measurements for stars, but their answer was always the same: the parallax angle p was found to be zero seconds of arc (0″). Such a result could be interpreted to mean that the stars are incredibly far away. Alternatively, it might mean that measurements of the position of a star made in July and in January are made from the exact same location in space; in other words, the measurement $p = 0″$ can be interpreted to mean that the Earth does not orbit the Sun. From the time of

Aristotle, in the fourth century BCE, until 1543, when Nicholas Copernicus suggested that the Earth orbits the Sun, the second choice was the preferred interpretation. Within a century of Copernicus presenting his heliocentric model, however, the first interpretation, that the stars are immensely distant, took precedence.

Parallax measurements are exceedingly difficult because the stars are very far away. The closest star, Proxima Centauri, has the largest parallax ($p = 0.772''$), barely three quarters of one second of arc. The difficulty becomes clear if we consider the degree of accuracy required. Prior to the invention of the telescope in the early seventeenth century, the most precise angular measurements made by astronomers for celestial objects (e.g., the angular distance between two stars) were made in the sixteenth century by the Dane Tycho Brahe. His most accurate measurements were within about one minute (sixty seconds) of arc. This meant that before astronomers would be able to measure the distances to the closest stars, they would have to improve the accuracy of their positional measurements by a factor of almost one hundred!

Even after the invention of the telescope, parallax measurements remained a formidable challenge. More than two centuries of technological improvements in the size and quality of telescopes had to occur before, in 1838, three astronomers, working independently, would make reputable measurements of the parallax angles of three different stars. From data collected over ninety-eight nights of observations, from 1837 into 1838, the German astronomer Friedrich Wilhelm Bessel announced in October of 1838 that he had measured the parallactic distance to the star 61 Cygni. With $p = 0.314''$, Bessel determined that 61 Cygni was about 3 parsecs away (the modern value is $p = 0.286''$, for which the distance is 3.5 parsecs). Two months later, the Scottish astronomer Thomas Henderson, having just recently taken up office as the Astronomer Royal for Scotland, reported that $p = 1.0''$ for Alpha Centauri, based on data he had obtained in 1832 and 1833 (the modern value is $p = 0.747''$, for which the distance is 1.34 parsecs). Next the German-born Russian astronomer Friedrich Georg Wilhelm von Struve calculated the parallax of Vega (Alpha Lyrae) as $p = 0.262''$ from measurements made on ninety-six nights over the course of the three years 1835–38 (the modern value is $p = 0.129''$, for which the distance is 7.75 parsecs).

In the 150 years following the breakthrough discoveries of Bessel, Henderson, and Struve, astronomers collectively invested enormous amounts of time and resources in efforts to measure the distances to stars. For most of those 150 years, progress was slow and painstaking. By 1878, astronomers had successfully used parallax measurements to determine distances to only 17 stars. By 1908, the number of stars whose distances were accurately measured through the method had reached 100; by 1952, the comprehensive and definitive *Yale Parallax Catalog* listed parallactic distances for 5,822 stars. Ground-based observing techniques had, by the 1950s, nearly reached their limit. The smallest parallax angles measureable using ground-based photographs and traditional measurement techniques are about 0.02 seconds of arc (0.02''), which corresponds to a distance of only about 50 parsecs. While there are several hundred billion stars in our galaxy and the volume of space within 50 parsecs of the Earth and Sun contains about 100,000 of them, most of these relatively nearby stars are exceedingly faint and therefore poor candidates for parallax measurements. Consequently, a tool that reached no further than 50 parsecs (a mere one million billion kilometers) put only a relatively small number of stars within our reach.

In 1980, recognizing the tremendous importance of parallax measurements for calibrating distances to all objects in the universe, the European Space Agency began planning the Hipparcos (High Precision Parallax Collecting Satellite) space astrometry mission. The satellite was launched in 1989, its four-year mission to measure the positions and parallax distances to more than 100,000 stars, with a limiting accuracy of 0.002 seconds of arc. Hipparcos was an overwhelming success, yielding as its final product a catalog containing the distances to over 120,000 stars with a limiting accuracy of only 0.001 seconds of arc. The success of the efforts described in this chapter cannot be overestimated. Virtually everything we know about stars and galaxies—and indeed about the universe—hinges on these essential and fundamental measurements.

◆ ◆ ◆ CHAPTER 6 ◆ ◆ ◆

Distances and Light

We further define the *absolute magnitude* (M) of a star, of which the parallax is π and the distance r, as the apparent magnitude which that star would have if it was transferred to a distance from the sun corresponding to a parallax of 0."1.

—J. C. Kapteyn, in "On the Luminosity of Fixed Stars," *Publications of the Astronomical Laboratory at Groningen* (1902)

Hipparchus and Magnitudes

In the second century BCE, the Greek astronomer Hipparchus compiled a catalog of about 850 stars and, like his mechanical successor the Hipparcos satellite, he noted the position and brightness of each star. Unlike Hipparcos, however, Hipparchus was unable to measure distances to stars; in fact he did not even try because he assumed, as did all astronomers of his day, that all stars were equally distant from the Earth. His brightness measurements, which he called *magnitudes*, were therefore comparisons of how bright the stars *appeared* in the night sky relative to each other.

According to Hipparchus, first magnitude stars were the brightest, second magnitude stars about two times fainter than first magnitude stars, and third magnitude stars two times fainter than those of the second magnitude. Hipparchus' faintest stars were designated "magnitude six" and were thought to be five magnitudes or thirty-two times fainter ($2 \times 2 \times 2 \times 2 \times 2 = 32$) than the brightest, first magnitude stars.

If, working with Hipparchus' magnitude scale, we were to discover a star two times brighter than the brightest star in Hipparchus' catalog, perhaps a star visible from South Africa but not from the Isle of Rhodes in the Mediterranean where Hipparchus worked, we would have to assign "magnitude zero" to this star. If we were to put the planet Venus on the same scale, its brightness would rank magnitude –5; the Sun would register about magnitude –27. Hipparchus' system has been modified by astronomers but is (unfortunately) still in fairly common use. In the modern magnitude system, stars that differ by one magnitude differ by a factor of about 2.512 in brightness, and a star whose magnitude is five less than that of another star is, by definition, one hundred times *brighter* than the second star.

The Inverse Square Law

Hipparchus thought his assigned magnitudes permitted him and others to make direct comparisons of the true brightnesses of stars, but he was of course wrong. His assigned magnitudes do permit us to compare how much light from each star reaches the Earth; but, because the amount of light reaching the Earth from a star depends on both the intrinsic brightness of the star and its distance from the Earth, we cannot make comparisons of the true brightnesses of stars unless we also know their distances from us.

Let's compare two light bulbs, each mounted in a small lamp that you can hold in your hands. We'll start by assuming that you do not know the strength of each bulb, but I know that bulb A is a 10-watt bulb and bulb B is a 90-watt bulb. I ask you to hold both lamps at arm's length, one in your left hand and the other in your right; bulb B will appear nine times brighter than bulb A. Since you know that both are equally distant from you, you can readily conclude that bulb B truly is intrinsically nine times brighter than bulb A.

Now, using these same two light bulbs, we'll put the 10-watt light bulb in a lamp located at a distance of exactly one meter from your eyes, but we'll put the 90-watt light bulb on a miniature flatbed railcar on a miniature train track. We'll slowly back that railcar away from you

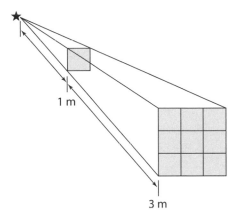

Figure 6.1. The inverse square law. When an object is at a distance of three meters from a light source, the light source must illuminate nine times more area than if the same object were at a distance of one meter. As a result, the light on each square is one-ninth as bright at a distance of three meters as at a distance of one meter.

until you declare that the two light bulbs appear equally bright. Since, in this case, you do know the intrinsic brightness of the two bulbs, you know that one of them, the intrinsically brighter one, is more distant than the other. You measure the distance to the 90-watt light bulb: three meters. At a distance (3 meters) that is three times greater than the distance (1 meter) to the fainter bulb, the bulb that is intrinsically nine times more luminous (90 watts) appears identical in apparent brightness to the fainter (10 watts) bulb.

Next we'll perform an experiment with two light bulbs of unknown intrinsic brightnesses. This time, however, you are not holding the lamps and you have no information about their distances from you, but when you measure their brightnesses you find that the two bulbs appear identical in brightness. In this experiment, how do you decide if the two light bulbs are identical in wattage and are located at equal distances from you or if one is intrinsically brighter but is placed at a greater distance? Without actually measuring the relative distances to these two bulbs (or stars), either explanation is equally plausible.

Light from a central point spreads out evenly in all directions. If we place a light source at the exact center of a giant spherical shell, that light source will illuminate with equal amounts of light every square meter of the inside surface of that sphere. We can calculate how much light each square meter of the inside surface of the sphere will receive by dividing the surface area of the sphere—which is proportional to the square of the radius of the sphere—by the wattage of the light source. If the sphere has a diameter of 4 meters and the source is a 50-watt light bulb, the intensity of light evenly illuminating the approximately 50 square meters of the inside surface of the sphere will be very nearly 1.0 watts per square meter. But what if we use this same 50-watt light bulb to illuminate the inside surface of a sphere that is ten times bigger, with a diameter of 40 meters? The level of illumination as those 50 watts spread out evenly across 5,000 square meters of the inside surface of the larger sphere will be one hundred times smaller, only 0.01 watts per square meter. In other words: the light intensity is one hundred times fainter when the distance to the light source is ten times greater. This is the *inverse square law* for light.

If I select two stars—call them Star A and Star B—that are identical in intrinsic brightness but one is ten times more distant, the more distant star will appear one hundred times fainter. If Star A is twice as far away as Star B, Star A would be four times fainter than Star B. On the other hand, if Stars A and B appear equally bright but I know that Star A is twice as far away as Star B, the more distant Star A must be intrinsically four times brighter than Star B in order to compensate for being twice as far away.

The Language of Astronomers:
Apparent Magnitude and Absolute Magnitude

In the language used by astronomers, *apparent magnitude* (for which we use the lower case *m*) indicates how bright a star appears to be in the nighttime sky. Hipparchus measured apparent magnitudes. *Absolute magnitudes* (for which we use the upper case M) are like the wattage measurements of the light bulbs; they indicate the intrinsic brightnesses of stars. Knowledge of absolute magnitudes requires that we measure both the apparent magnitudes and the distances of stars. Assessing the apparent magnitude of a star is a straightforward business. You observe the brightness of a star (with or without a telescope) and make a judgment (with your eyes or with a more sophisticated measuring device, such as camera film or a digital camera) as to its apparent brightness.

Absolute magnitudes, however, which are the magnitudes that truly permit us to compare one star with another, are much more difficult to measure because, for these, we need to know both the apparent magnitudes of the stars and their distances from us.

Absolute magnitudes were defined in 1902 by the great Dutch astronomer Jacobus Cornelius Kapteyn as the brightnesses stars would have if they happened to be at distances of exactly 10 parsecs. The International Astronomical Union institutionalized this definition at its very first meeting in 1922, and astronomers ever since have used absolute magnitudes when they wish to discuss and compare brightnesses of stars. Of course, no stars are actually at a distance of exactly 10 parsecs, so the absolute magnitude of any star is calculated by applying the inverse square law for light to the measurements of its distance and apparent magnitude. With those measured quantities, we calculate how bright that star would be if we moved it to a distance of 10 parsecs.

How would an astronomer obtain the absolute magnitude for a particular star? First, he measures the apparent magnitude of the star and finds that it has an apparent magnitude of +10 ($m = +10$). He then measures the parallax of the star and finds $p = 0.01''$. From the parallax, he computes the distance to the star: 100 parsecs ($d = 100$ pc). Since he wants to compare the brightness of this star to other stars, he then asks, What is the absolute magnitude of that star? The absolute magnitude is the brightness the star would have were it at a distance of only 10 parsecs. So he asks, What would happen (in my thought experiment) to this star if I moved it in from a distance of 100 parsecs to a distance of only 10 parsecs? Our astronomer's answer: the star would be ten times closer; by the inverse square law, ten times closer would make it 100 times brighter; and according to the definition of the magnitude system, 100 times brighter is the same as five magnitudes smaller, so this star would have an absolute magnitude of +5 ($M = +5$). While the apparent magnitude of the Sun is −27, the absolute magnitude of the Sun is about +5, so we conclude that this particular star with an apparent magnitude of +10 is comparable in brightness to the Sun.

The Hipparcos satellite measured both the apparent magnitudes and parallaxes (i.e., distances) for over 100,000 stars within its reach. What did Hipparcos find? In the neighborhood of the Sun, the very brightest star, in terms of absolute magnitudes, is Beta Orionis, with an absolute magnitude of –6.69. The very faintest stars have absolute magnitudes of over +13. This information tells us that in the solar neighborhood, the brightest star is about 11.5 magnitudes brighter than the Sun, which makes it about 40,000 times brighter than the Sun; the faintest star is about 8 magnitudes fainter than the Sun, making it almost 2,000 times fainter than the Sun. If we compare the brightest star to the faintest star, we find that the former is about 80 million times brighter than the latter. Suddenly, with a set of careful measurements of only two stellar parameters, the apparent magnitude and the parallax angle, we find ourselves doing astrophysics. We know that the range of brightnesses of stars is phenomenally large and that the Sun is a middle-of-the-pack star, much brighter than the faintest stars, much fainter than the brightest. Our ability to measure and compare the absolute brightnesses of stars will prove to be the critical tool that will allow us finally to step out to the edge of the universe and to determine its age.

···CHAPTER 7···

All Stars Are Not the Same

If, on the contrary, two stars should really be situated very near each other, and at the same time, so far insulated as not to be materially affected by the attractions of neighboring stars, they will then compose a separate system, and remain united by the bond of their own mutual gravitation towards each other. This should be called a real double star; and any two stars that are mutually connected, form the binary sidereal system which we are now to consider.

—Sir William Herschel, in "Catalogue of 500 New Nebulae, Nebulous Stars, Planetary Nebulae, and Clusters of Stars; With Remarks on the Construction of the Heavens," *Philosophical Transactions of the Royal Society of London* (1802)

Two millennia ago, Hipparchus made the assumption that all stars are the same distance from Earth and therefore differ in brightness because some are intrinsically brighter than others. It was a reasonable assumption for his time, but by the eighteenth century it was no longer tenable. Aristotle's physics and his geocentric cosmology had been replaced by Newtonian physics and by the heliocentric cosmology of Copernicus. Astronomers would remain unable to measure the distances to any stars until the fourth decade of the nineteenth century, but already they were in universal agreement that the stars in the heavens were at many varied distances from the Earth. Consequently, but without good reason, eighteenth-century astronomers turned Hipparchus upside down and decided that all stars are identical in all their properties except for their distances. Therefore, bright stars were understood to be bright simply and only because they were closer than faint stars.

Not Equally Bright

William Herschel, professional musician and self-taught astronomer, was responsible for the most important astronomical research done in the eighteenth century, and he made almost all of his discoveries with a 19-inch-diameter, 20-foot-long telescope that he built in his own garden. Though Herschel is best remembered for discovering the planet Uranus, he actually devoted most of his attention to the study of stars. Initially, Herschel accepted the prevailing premise that all stars were identical in all their intrinsic properties. He also assumed that stars were equally spaced one from another and that that mean distance was the (unknown), but presumably identical, distance from the Sun to either Sirius or Arcturus, the two brightest stars (in appearance) that he could see in his English nighttime sky and therefore, following his logic, the two closest stars to the Sun.

As part of his research plan, Herschel devoted much effort to measuring the positions of stars that appeared to be near one another in the sky. He assumed such *double stars* were in fact two stars that happened to lie in similar directions in the sky, but that one was in fact much more distant than the other. Starting from the premise that the brighter star in each pair must be much closer than the fainter star, he worked to determine the parallax of the brighter and presumably closer star by watching for the closer star to move with respect to the more distant star. While he never succeeded in measuring the parallax for even one star, he did discover that the close association of a faint star with a bright star occurred with a frequency that greatly surpassed the number of such associations one would expect if stars were distributed uniformly in the heavens. He also discovered, by the early 1800s and after twenty years of observing double stars, that the angle in the sky between several faint-bright star pairs had changed in a continuous and predictable way. The changing positions of the stars in these pairs revealed that these pairs of stars were what he called binary star systems—Herschel coined the term *binary star* in 1802—which consisted of two stars orbiting a common center, obeying Newton's law of gravity, and lying at essentially identical distances from the Earth and Sun. This discovery was momentous, not only because Herschel had discovered the existence of a new class of stars, but also because the existence of both faint and bright stars in binary systems demonstrated conclusively that all stars are not identical. Some are intrinsically faint, while others are intrinsically bright. At the dawn of the nineteenth century, astronomers were forced to recognize that stars differed in both intrinsic brightness and distance.

Not Identical in Color

As early as the second century CE, the great Greek astronomer Ptolemy reported that six stars—known to us as Aldeberan, Antares, Arcturus, Betelgeuse, Pollux, and Sirius—were yellowish in color, while other stars were white. In the intellectual context of his time, the yellow coloring was presumed to be caused by starlight passing through Earth's atmosphere and not by color differences intrinsic to the stars.

In the late 1770s, William Herschel also took notice of the differences in star colors and, in 1798, conducted a study of six stars which revealed that Aldeberan, Arcturus, and Betelgeuse were more red and orange than were Procyon, Sirius, and Vega, which more uniformly showed all the colors from red to violet. Unwilling to conclude that these colors were intrinsic properties of the stars, he speculated, incorrectly, that the colors indicated something about the motion of the stars. By Herschel's death in 1822, no satisfactory explanation for the color differences had yet been found. It was not until the following decade that Struve, noting widely contrasting colors—red, blue, and green—in binary star systems, showed conclusively that these color differences were intrinsic to the stars and were due neither to atmospheric effects nor to their motions (in the 1840s, Christian Doppler would propose a method by which very small color differences among some stars might be explained by their motions; see Chapter Ten). By the 1830s then, astronomers recognized that stars differed in both brightness and color.

Distinct Spectra

The 19 February 1672 issue of *Philosophical Transactions* includes a letter from Isaac Newton on the colors of light. This paper, which established his reputation as a natural philosopher, presented his work on optics in which he established that white light is composed of a spectrum of colors, ranging from violet to red.

In 1802 the English chemist William Wollaston and in 1814 the German glassmaker Joseph Fraunhofer independently discovered dark lines in the spectrum of the Sun. Wollaston noted seven such lines; Fraunhofer identified hundreds, which he was convinced were intrin-

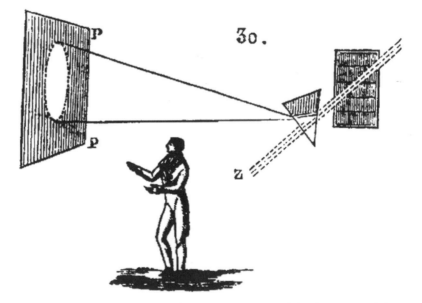

Figure 7.1. Newton's experiment with white light, in which he demonstrated that white light from the Sun can be dispersed into a rainbow of colors, as presented in a sketch redrawn from Voltaire's *Eléments de la Philosophie de Newton*, published in 1738.

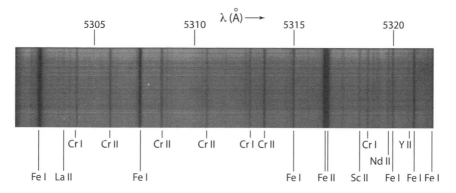

Figure 7.2. A modern spectrum showing the kinds of dark lines originally seen by Fraunhofer and Wollaston in the early nineteenth century. The labels at the top indicate the wavelength of light (in units of angstroms; one angstrom is one ten-billionth of a meter), the labels at the bottom which elements are responsible for which lines. Image courtesy of E. C. Olson, Mount Wilson Observatory.

sic to the Sun. Three years after his initial discovery, Fraunhofer published his discovery in his memoir "The Determination of the Refractive and the Dispersive Power of Different Kinds of Glass" that the spectra of Sirius also showed three dark lines, one in the green and two in the red, "which appear to have no connection to those of sunlight." He continued, "In the spectra of other fixed stars of the first magnitude, one can recognize bands (dark lines); yet these stars, with respect to these bands, seem to differ among themselves."

And so it was that by 1840, William Herschel's identical stars had given way to stars that differed in brightness, color, and spectral features. Astronomy was about to give way to astrophysics.

What Is Light?

In order for us to follow the important astrophysical discoveries of the rest of the nineteenth century and beyond, we will need to understand several properties of light. Light, of course, is absolutely fundamental to astronomers: it is what we measure when we observe celestial objects. So, in this next section we will take a brief side-trip to explore the basic physics of light and to explain how its properties contribute to our learning about the stars.

Light is energy in the process of moving through space. As it travels it sometimes bounces (reflects) off surfaces, just as a tennis ball bounces off the ground; in other words, it behaves like a solid particle. At other times and in other circumstances—for instance when light passes through a slit or bends around a corner—this traveling packet of energy we call light behaves like a wave. Physicists invented the name *photon* for these light packages that travel through space sometimes exhibiting particle-like and at other times wave-like properties.

The speed of light in a vacuum (a volume of space in which there are no particles of mass) is about 300,000 kilometers per second. Light travels slightly slower through air or glass or water, and in a particular medium, longer wavelength photons will travel slightly faster than shorter wavelength photons.

Photons are characterized by their wavelength, frequency, and energy. The wavelength is the distance from one wave crest to the next. Given that all light waves travel at identical speeds in a vacuum, fewer waves of light with a large wavelength than waves with a shorter wavelength will pass a given point in a fixed amount of time. The number of waves passing a fixed point in a single second is the frequency of the light (measured in units of waves per second or cycles per second). The wavelength and frequency are inversely related to each other, such that the wavelength multiplied by the frequency is equal to the speed of light. Since the speed of light in a vacuum is a constant, a photon with a large wavelength has a small frequency and one with a small wavelength has a large frequency. The energy carried through space by each photon is directly proportional to its frequency and inversely proportional to its wavelength. Thus, higher frequency (shorter wavelength) photons carry more energy than lower frequency (longer wavelength) photons.

When we discern colors, we are actually detecting and measuring different wavelengths of light. Our eyes happen to be moderately efficient systems for detecting certain wavelengths, the ones that produce the rainbow of colors between violet and red on the light spectrum. Scientists long assumed that these were the only possible colors. In 1800, however, William Herschel showed that light will heat up thermometers and that a thermometer set just beyond the red end of a visible light spectrum, where our eyes see absolutely no light, also will heat up. Herschel had discovered that the spectrum of light does not end with the longest wavelength of light perceptible to human eyes; instead, the spectrum continues beyond the red, beyond where our eyes can see color, into a region known as the *infrared*. After learning about Herschel's experiment, the German chemist Johann Wilhelm Ritter used the same experimental technique in 1801 to discover light of wavelengths shorter than the shortest (violet) light that our eyes can detect, the region of the spectrum that is known to us as *ultraviolet light*.

We now understand that the electromagnetic spectrum extends from the very highest energy photons, known as *gamma rays*, to the slightly less energetic *X-rays*, then to *far ultraviolet* ("far" from visible), *near ultraviolet*, and *visible light*, on to *near infrared* and *far infrared light*, *microwaves*, and then finally at its very low energy (and long wavelength) end, to *radio waves*. Since our eyes are unable to detect colors outside the narrow range of visible light, we need other instruments to detect other kinds of electromagnetic waves. Human bones, for example, are good X-ray detectors: this is because the density of a bone stops X-rays so that the picture taken by the radiologist reveals a negative image of the bone. Melanin molecules in human skin cells are excellent detectors of ultraviolet light, and water molecules are very efficient at detecting infrared photons and microwaves. Of course, human bone and skin as well as water molecules are of little use for making quantitative measurements of the brightnesses of stars and galaxies; so astronomers have designed and built a broad range of devices that are capable of detecting light from astrophysical sources across the entire electromagnetic spectrum.

For reasons similar to those that explain why different materials are more or less effective at detecting different wavelengths of light—for example, the composition, density and temperature of those materials—astrophysical objects can look very different when observed at different wavelengths.

When viewed in X-rays, the Crab Nebula looks like a spinning disk with a jet emerging from that disk, while in ultraviolet light it looks

Figure 7.3. Top, the Milky Way seen in visible light. There are very bright regions due to the accumulated light from millions of stars, as well as numerous patches of darkness where dusty, interstellar clouds block the light of the stars. Bottom, the Milky Way as seen in infrared light. The overall bright glow is produced mainly by heated clouds of interstellar dust. Images courtesy of Axel Mellinger (top), E. L. Wright, The COBE Project, DIRBE, and NASA (bottom).

like a bubble filled with glowing filaments. By using telescopes designed to measure different wavelengths of light, astronomers can learn about the physical processes in stars, galaxies, and interstellar space that produce those kinds of light. By studying a single object with myriad kinds of telescopes and at different wavelengths they can learn about the widely varying astrophysical phenomena that occur far beyond the Earth.

Colors of Light

With our eyes, we perceive that objects have distinct colors: blades of grass are green, blooming roses are red and yellow, ripe blueberries are blue, hot embers are black or red or orange or even white, a flame might be yellow or orange or blue, and the Sun is yellow. Objects have different colors for one of a few reasons: either they emit light of a distinct color; or they reflect light of only that color; or they may emit light of all colors but more of one color than of the others so that one color is dominant.

As Newton showed, white sunlight is composed of all the colors of the visible rainbow. When white sunlight reflects off a carrot, the chemicals in the carrot absorb all the incident colors of sunlight except orange. The orange light—and only the orange light—is not absorbed but reflected, and the carrot therefore appears orange. A ripe lemon absorbs all colors except yellow and reflects yellow. Green leaves absorb all colors except green and reflect green.

The light the Earth receives from the Sun is known as a *continuous spectrum* of visible light because it includes all the colors our eyes can see. The spectrum reflected from a leaf in summer is continuous across all the shades of green but is no longer continuous across all colors because most of the other colors have been absorbed.

When white sunlight passes through a gas, such as the Earth's atmosphere, most of the continuous spectrum is transmitted through the gas; however, in most situations, a few specific colors, perhaps a certain shade of red and certain shade of yellow, will be filtered out of the originally continuous spectrum. The resulting spectrum is known as an *absorption spectrum*.

The Sun is yellow because, although it emits light of all colors, it emits more yellow light than light of any other color. A very hot piece of wood emits light in all the colors to which our eyes are sensitive; the blend of all these colors appears white. That same piece of wood appears red if it is much cooler because, at the lower temperature, it emits less violet and green and blue so that the red light is dominant. Any object that is sufficiently dense (such that the particles that make up the object touch or collide often) and any sufficiently large object, even of very low density, emits a continuous light spectrum; the amount of light emitted at each and every color (across the entire electromagnetic spectrum, from gamma rays to radio waves) depends only on the temperature of the object. The light from such an object is known as *thermal* or *blackbody* radiation.

A Star Thermometer

A blackbody is an ideal object that emits electromagnetic radiation in a way that depends solely on the temperature of the object (no matter how high or low the temperature). By applying the mathematical physics for blackbodies that was worked out a century ago by the German physicist Max Planck, it is possible to describe exactly how much energy is emitted at each and every possible wavelength or frequency of light for an object of a given temperature. A plot of the energy emitted by an object as a function of wavelength or frequency is known as a *blackbody spectrum* or *Planck spectrum*. On such a plot, and in fact in all work done by astronomers, temperatures are measured on the Kelvin scale. Water boils at sea level at a temperature of 373 K (100 °C; 212 °F) and freezes at 273 K (0 °C; 32 °F).

Blackbody spectra have a number of important characteristics that are well worth noting: as defined, blackbodies emit light at *all* wavelengths, from gamma rays to radio waves; the amount of light emitted by a blackbody at each wavelength increases rapidly from the shortest wavelengths to a peak wavelength and then decreases, though less rapidly, toward the longest wavelengths. The wavelength at which a blackbody emits the most light is shorter for hotter objects, longer for colder objects; and, from identical areas of their surfaces, a hotter blackbody emits more light or energy than a colder blackbody. Extremely hot objects (millions of degrees), such as the disks around black holes, emit most of their light as X-rays. Objects with temperatures of tens of thousands of degrees (the hottest stars) emit most of their light in the ultraviolet range. Objects with temperatures of thousands of degrees (stars like the Sun) emit mostly visible light. Objects with temperatures of hundreds of degrees (such as you or me or the Earth) emit most effectively in the infrared range (though such objects do also emit extremely faint amounts of red light that can be collected and amplified by night-vision goggles).

As we have noted, the wavelength at which a blackbody emits the

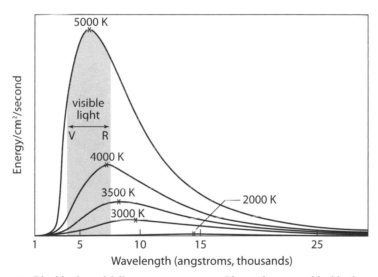

Figure 7.4. Blackbodies of different temperatures. Objects known as blackbodies emit energy in the form of light at all wavelengths, from gamma and X-rays (left) to radio waves (right). At a given temperature, a blackbody emits more energy per second at one wavelength than any other wavelength. The wavelength at which a blackbody emits the most energy depends only on the temperature of the blackbody. This illustration shows blackbody curves for blackbodies at temperatures of 5,000 K (close to the temperature of the Sun; most of the light is emitted as visible light), 4,000 K, 3,500 K, 3,000 K, and 2,000 K (most of the light emitted is infrared).

maximum amount of light is determined only by the temperature of the object; this relationship is known as Wien's law. Bigger objects emit more total light than smaller objects of the same temperature, but the size of the object does not affect the wavelength of light at which the object emits most effectively. Wien's law tell us that if we can measure the light from an object at enough wavelengths to determine two things—first, that it emits light like a blackbody; and then the wavelength at which it emits the most light (its color)—we can use this information to calculate the temperature of that object.

Most stars emit light approximately as blackbodies. In practice, astronomers can measure the amount of light emitted by a star with a violet filter (i.e., measuring only the violet light), then with a blue filter, then with green, yellow, orange, and red filters. Moving outside the range of visible light, they can further observe the star with filters in the X-ray and ultraviolet and infrared and radio ranges. In the end they can chart the intensity of the star's light across the entire wavelength range of the electromagnetic spectrum. By comparing the star's electromagnetic profile with the known profiles for blackbodies of different temperatures, astronomers are then able to determine the temperature of the star. In effect, because stars emit light like blackbodies, all that astronomers need do is determine the wavelength at which a star emits more light than at any other wavelength; that measurement immediately yields the temperature of that star.

Wien's law is an incredibly powerful tool in the astronomer's arsenal; it is no less than a thermometer for stars. It explains why stars have different colors—yellow indicates a star with a temperature comparable to our Sun's (about 6,000 degrees), red a cooler star (a few thousand degrees), and blue a much hotter star (around 20 thousand degrees).

Inside the Atom: Light Interacting with Matter

When light interacts with matter, three things can happen: the light can be reflected by, absorbed by, or transmitted through the matter. In order to explain which of these three modes of interaction will occur

in a given situation, we need to understand something about the structure of an atom.

Atoms are composed of three fundamental building blocks: protons,
neutrons, and electrons. Protons and neutrons are confined to the
atomic nucleus while electrons form a cloud that surrounds the nucleus. Electrons gain and lose energy according to the rules of quantum
mechanics, which govern the behavior of these and all other subatomic particles. An electron can gain energy by absorbing a photon and
lose energy by emitting a photon. If an electron is *free*—that is, not
bound in orbit around a nucleus—it is permitted to absorb or emit a
photon of any energy; however, if the electron is in a bound orbit within an atom, it is only able to absorb (or emit) photons that contain the
exact energies that would allow the electron to jump upwards (or drop
downwards) to distinct, higher (or lower) energy orbits around the
nucleus or to absorb any photon with enough energy to completely free
it from its bound state. If, for any reason, an electron drops from a
higher to a lower energy orbit, it emits a photon with an exact energy
that corresponds to the energy difference between the two orbits. Since
a photon with that energy has a unique wavelength (or frequency),
this electron emits a photon with the exact color that is associated
with that wavelength.

For every element, the rules of quantum mechanics—in which each
electron must be thought of as a spread-out wave rather than as a particle at a fixed position—govern the possible energy levels that its electrons may occupy. Each energy level is called an *orbital*, which is a region of space surrounding the nucleus in which an electron is likely to
be found. Rather than a series of concentric rings around a bull's eye
the orbitals are like bubble-gum bubbles surrounding the nucleus, and
each electron is a wave that exists within one such bubble. The sizes
and shapes and distances from the nucleus of the energy orbitals are
determined by the number of protons and neutrons in the atomic nucleus. For example, all atoms of carbon-12 in the universe are identical—with six protons and six neutrons in the nucleus—and the rules
of quantum mechanics are identical for every one of these ^{12}C atoms.
These rules determine a distinct set of energy levels for the electrons in
^{12}C atoms that will be different from the orbitals for any other atomic

species, even for other isotopes of carbon, and that will change if the temperature of the carbon atoms changes. At any temperature, each of the six electrons is required to have an energy that places it somewhere within an orbital that occupies a distinct volume of space at a particular distance from the nucleus —and these orbitals, separated by well-defined energy differences, are the only locations in which the six electrons in a ^{12}C atom may be found. In a ^{13}C atom, the quantum mechanical rules that define the locations of the orbitals of the six electrons will generate slightly different orbitals than will the rules that define the energy levels of the six electrons in a ^{12}C atom with the same temperature; consequently, the amount of energy needed to move from one orbital to the next in the ^{13}C atom will be very slightly different from the spacings of the energy levels in the ^{12}C atom. Since each energy difference corresponds to a unique wavelength of light, every isotope of every element is permitted to absorb or emit only the specific colors of light that correspond to the energies that are associated with the differences between energy levels of the electrons in that isotopic species. The permitted colors for ^{12}C are like a set of fingerprints, a spectral signature that distinguishes ^{12}C from ^{13}C or ^{238}U or any other element. Oxygen atoms have a set of permitted colors that are unique to oxygen, iron atoms have a set of permitted colors that are unique to iron, and every other atom and every isotope of every atom, and in fact every molecule, has a quantum-mechanically determined spectral signature.

If we had a tube of gas (or a lamp) filled only with sodium atoms and if we heated that tube or ran an electric current through it, the electrons in the sodium atoms would absorb energy, which would propel them to higher energy orbits. These higher energy orbits are unstable, so the electrons would then emit photons and drop down to lower energy levels, thereby cooling the gas. When the sodium gas emits photons, we say that the sodium gas is glowing; this glowing gas emits light at only a few distinct, well-separated wavelengths. We call the light from such a gas an *emission spectrum* made up of distinct emission lines.

If a cold cloud of sodium gas were located between us and a distant source of continuous light, the photons from the continuous spectrum (from a star for example) would have to pass through the sodium cloud in order to reach our eyes. Because each photon carries energy, these photons in the continuous spectrum become a heat source for the sodium gas. But the sodium gas can absorb light at only a few distinct wavelengths corresponding to the energy differences between permitted electron levels for sodium atoms. As a result, the light that enters the sodium cloud differs from the light that exits the cloud. Entering light contains all colors, continuously across all wavelengths, but the light that emerges from the cloud is missing light at those few distinct wavelengths that the electrons in the sodium atoms are able to absorb. The result is a continuous spectrum with a few distinct colors missing. We call this spectrum an *absorption spectrum*.

This phenomenon of the absorption of a few distinct colors of light by a cold cloud of material that lies between the continuous light source and our telescopes is exactly what Fraunhofer saw when he observed the Sun, Sirius, and other bright stars. In a star, the source of continuous light, the layer that emits most of the light, is called the photosphere. The gas cloud in between us and the light source in the Sun is composed of the tenuous outermost layers of the Sun's atmosphere, in which the gas is slightly cooler and much less dense than the gas that makes up the Sun's photosphere. Light from the solar photosphere passes through the outermost layers of the Sun in order to escape from the Sun; in doing so, some of that light is absorbed by the elements in these uppermost layers. The resulting spectrum is missing light at discrete wavelengths; the places where the light is missing are the black lines that Fraunhofer (and Wollaston) saw. The outermost layers of the atmospheres of Sirius and other stars reveal elemental signatures that are slightly different from the Sun's. These differences are due only in small part to actual differences in the chemical compositions of these stars' atmospheres. The most important reason for the discrepancies in their spectral signatures is that the gases in the outer atmospheres of Sirius and of the other stars Fraunhofer observed are at slightly different temperatures than the gases in the outermost layers of the Sun. The temperature of the gas in turn determines the photon energies (manifest as spectral lines) at which the gas can absorb. Since, for each star, the set of dark lines fingerprint both the elements that are present and the temperature of those elements in the absorbing outer layer of the atmosphere of that star, the spectra of Sirius and of other bright stars show sets of dark lines that differ from the dark lines seen in the solar spectrum.

The Stefan-Boltzmann law

Experience has taught us that if we have two objects of different sizes but with the same temperature, the larger of the two emits more energy. For example, if we wish to boil a pot of water on an electric stove, the pot will boil more quickly if we place it on a large heating element rather than on a small one, assuming both heating elements are at the same temperature setting. Experience, however, also has taught us that the smaller electric heating element with the temperature control set on "high" can heat up our pot of water as quickly as can the larger one when the temperature control for the larger one is set on "low."

These two articles of common sense have been codified by physicists as the Stefan-Boltzmann law: the total energy released per second by an object (known as the object's *luminosity*) depends both on its emitting surface area and its temperature. If the object gets bigger and thus has a larger surface from which to emit light, the luminosity increases (if we double the surface area, we double the luminosity); however, if the temperature increases, the luminosity goes up much more dramatically (if we double the temperature, the luminosity becomes 16 times greater).

Astronomers keep the Stefan-Boltzmann law along with Wien's law and the inverse square law for light ready to hand in their toolkits. With the temperature of a star measured with Wien's law and the luminosity of the star calculated from the inverse square law for light (which in turn requires measurements of the parallax [distance] and apparent luminosity of the star), astronomers can determine the total surface area (or radius) of that star. This is quite an impressive result to achieve with such a relatively small arsenal of tools.

We can now return to the astronomical work of the nineteenth century and learn how astronomers came to understand the basic astrophysics of stars. That understanding will provide us with additional tools to use in measuring distances to more distant objects in the universe. Then we will be able to unravel the great nineteenth-century mystery of the spiral nebulae. And finally we will use what the nebulae teach us to take several dramatic leaps toward the answer to our question: How old is the universe?

CHAPTER 8

Giant and Dwarf Stars

One can predict the real brightness of a dwarf star from a knowledge of its spectrum alone.

—Henry Norris Russell, in "'Giant' and 'Dwarf'" Stars," his June 13, 1913 address to the British Astronomical Society, published in *The Observatory*, 1913

For four decades, from the 1820s through the 1850s, a broad spectrum of the scientific community—astronomers, physicists, chemists, glassmakers, and even photographers—turned their attention to spectra. They studied the spectra of celestial objects, of chemical elements, of enclosed vapors in laboratory settings, and of the Earth's atmosphere. From their measurements and experiments, they generated the knowledge that would lead, by the end of the century, to an effective methodology for classifying the stars. By the early twentieth century, astronomers would go a step further, using these spectral classifications, in combination with other measurements of stars, to unveil the types and life cycles of stars and to make possible measurements that would in the end answer profound questions about the nature of the universe and indeed about its age.

One of the very earliest observations, and a foundational one, was made by Fraunhofer, who discovered that the spectrum of the Moon showed the same dark lines as the solar spectrum. He also recognized several of the darkest of these lines in the spectra of Venus and Mars. Fraunhofer never found an explanation for any of these lines, nor for the different dark lines that he saw in the spectra of some stars, but others soon would.

In the 1820s, John Herschel, William Herschel's son, who would become nearly as famous an astronomer as his father, demonstrated that when metallic salts are heated they do not emit light in a continuous spectrum but, instead, emit light in distinct, bright spectral lines that could be used to identify the heated materials. His contemporary William Henry Fox Talbot, one of the inventors of photography, similarly showed that the known elements, when heated, have unique sets of bright spectral lines. Soon thereafter, in 1833, the Scotsman David Brewster, who is most famous for inventing the kaleidoscope, showed that some of the dark lines and bands in the solar spectrum are due to absorption by the Earth's atmosphere. Brewster noted that some of his lines and bands changed depending on the proximity of the Sun to the horizon, becoming broader and darker the longer the path the sunlight had to travel through the Earth's atmosphere, while others appeared unchanged by Earth's atmosphere and therefore must be intrinsic to the Sun.

In 1849, Jean Bernard Leon Foucault, the French physicist famed for his invention of his eponymous pendulum and for taking the first photograph of the Sun in 1845, discovered that he could make certain dark lines in the solar spectrum darker by passing sunlight through sodium vapor. These particular dark lines, which Fraunhofer dubbed "D lines," also showed up as bright lines in light from a sodium lamp in Fraunhofer's original work. Heated sodium vapor, Foucault concluded, emits light that generates these bright spectral lines; in contrast, cool sodium vapor absorbs light at these same wavelengths and generates dark spectral lines when the light begins elsewhere and passes through the cool sodium vapor. At this time in the middle of the nineteenth century, astronomers and physicists had reached the tipping point in understanding the phenomena of bright and dark spectral lines.

Physicist Gustav Robert Kirchhoff and chemist Robert Bunsen, while working together in Heidelberg in the 1850s, developed improved methods for purifying substances. With these purified materials, they demonstrated definitively what Fox Talbot's work had already suggested, that every element produces a unique set of spectral lines, and what Foucault had discovered, that an element can both emit and absorb light at those same wavelengths. Emission of light shows up as

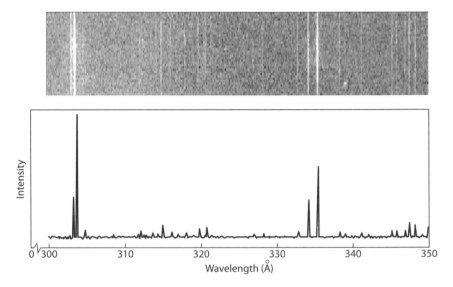

Figure 8.1. Two views of an ultraviolet spectrum of the Sun. Top, the photographic image of the spectrum, revealing the bright lines that mark wavelengths at which ultraviolet light, between the wavelengths of 300 and 350 angstroms, emerges from the Sun. Bottom, a plot of the intensity or brightness of the spectrum across the same wavelength range. Image courtesy of NASA.

discrete spectral lines when the element is in the form of a hot, rarefied vapor; absorption of light produces discrete spectral lines when the element is in the form of a cool vapor and lies in between a light source and the viewer; and emission of a continuous spectrum (for which Kirchhoff coined the term *blackbody spectrum*) occurs when an object is hot and dense. Armed with this understanding of light, Kirchhoff and Bunsen were able to identify specific elements in the spectrum of the Sun. In so doing, they inaugurated the new era of stellar spectroscopy, which would dominate astronomy for the next half century.

A Transition: From Naming Stars to Classifying Stars

Some of the brightest stars in our heavens are known to us by given names —Polaris, Sirius, Antares, Rigel. Being able to point to a star and call it by its personal name may give us a comfortable sense of familiarity, but it does very little to advance our astronomical knowledge.

The first attempt to classify stars using what were thought to be their actual characteristics was undertaken by the German astronomer Johann Bayer. Bayer, in his book *Uranometria* (1603), invented a naming scheme that combined a lower case Greek letter with a Latin constellation name (in the genitive form): "Alpha Canis Majoris" (also known as Sirius) was the designation for the brightest star in the constellation Canis Major, "Beta Orionis" (also known as Rigel) designated the second brightest star in Orion. In this way Bayer classified more than 1,300 stars in forty-eight different constellations. Regrettably, however, his formula relied on criteria that had nothing to do with the astrophysics of the stars and so produced a classification system that was really little more than a new naming scheme. The stars in a single constellation usually include faint, nearby stars and bright, more distant stars, few if any of which have any physical relationship to any of the others; furthermore, the brightest star might be intrinsically faint, but nearby, while the faintest star might be intrinsically bright, but distant. Bayer's scheme labeled more stars than the older personalized scheme, but it did not contribute to a better understanding of the stars themselves.

A century later, the English astronomer John Flamsteed created a similar classification scheme using, like Bayer, the Latin genitive form of the constellation name, but combining it with Arabic numerals, beginning with the westernmost star in that constellation: Sirius became "9 Canis Majoris," and Rigel "19 Orionis." Like Bayer's method, this naming scheme works well for identifying a star by its apparent relative brightness within a region of the sky or by its position in the sky, but it provides no information about the star itself. Is "3 Lyrae" the brightest star in Lyra because it is close to the Sun or because it is intrinsically bright?

The understanding of spectral lines that emerged from the work of Kirchhoff and Bunsen offered an improved way to name and classify stars; rather than by their positions or their apparent brightnesses, astronomers could classify stars on the basis of their colors and spectra, properties intrinsic to the stars themselves. From 1863 to 1868, the

Jesuit priest Angelo Secchi, who in the 1850s rebuilt the observatory at the Collegio Romano in Rome, examined the spectra of over 4,000 stars and classified them into four broad categories. About 50 percent of Secchi's stars were white or blue stars (like Sirius and Vega) with spectra dominated by four absorption lines attributed to hydrogen atoms. He called these stars of the "1st type." Stars of the "2nd type" were yellow (like Capella and Aldebaran) with spectra like the Sun's. The orange-red "3rd-type stars" (like Betelgeuse and Antares) had no hydrogen lines, and the faint, red "4th-type" stars (like R Cygni) had dark bands of lines attributed to hydrocarbons.

Secchi's four spectral types, which classify stars according to their intrinsic light-emitting properties rather than by their locations in the sky, represented a major advance in the study of these celestial objects. But there was of course much more to be discovered. Astronomers of the late nineteenth century believed, for example that the four categories were stages in a predictable pattern of aging. Stars, they presumed, were born as large, hot, blue objects (1st type), then cooled and faded with time, turning first from blue to yellow (2nd type), then to orange-red (3rd type), and finally to red (4th type) as they used up their supplies of internal heat. Finally, they had a method of classifying stars that could inform them directly about their astrophysical properties. If, from a spectrum alone, an astronomer could conclude "that's a 4th type star," then all astronomers would instantly know that this particular star was very old. Unfortunately, this understanding of these four spectral types proved completely wrong. 4th-type stars are not necessarily old, nor are 2nd-type stars necessarily middle-aged. It was not long before other astronomers began to improve on Secchi's work.

Spectral Photography

Henry Draper, by profession a physician and eventually Dean of Medicine at New York University, was an amateur astronomer who is famous for having obtained the first photograph of the Orion Nebula in 1880. Eight years earlier, he had obtained the first photograph of a stellar spectrum, that of Vega, and by 1882, the year of his death, his col-

Figure 8.2. Left, undated photograph of Henry Draper. Right, undated photograph of Draper at his telescope. Left image courtesy of the Draper Family Collection, Archives Center, National Museum of American History, Smithsonian Institution; right image courtesy of the Photographic History Collection, Division of Information Technology and Communications, National Museum of American History, Smithsonian Institution

lection would include the photographic spectra of fifty stars. In 1886 his widow donated his equipment and provided an endowment to the Harvard College Observatory to establish the Henry Draper Memorial Fund, with the intent that these funds support additional work along the lines established by her husband.

Edward Pickering, the Director of the Harvard College Observatory from 1879 until 1919, did just that. With a telescope that could image a large area on the sky, a prism that converted the light of every star into a tiny image of its spectrum, and the ability to capture this image of multiple stellar spectra on a single photographic plate, Pickering and his team of astronomers generated a photographic archive of spectral data for more than 10,000 stars that by 1890 would serve as the basis for the *Draper Catalogue of Stellar Spectra*.

In order to analyze the thousands of spectra on the photographic

Figure 8.3. Left, Edward Pickering. Right, room full of women "computers" at Harvard College Observatory, 1892. Fleming is seen standing in the middle of the group. Images courtesy of Harvard College Observatory photo collection.

plates, Pickering hired a number of women (whom he paid a fraction of the salary of men doing the same work) to act as his "computers." The first of these was Williamina Fleming. Originally hired as Pickering's housekeeper, she made the transition to the clerical staff of the Harvard Observatory in 1881 and was soon assigned the task of studying the stellar spectra that were arriving at the Observatory as the first fruits of the Draper family bequest.

Fleming devised a system of assigning to each star a letter that corresponded to how much hydrogen could be observed in its spectrum. She used this scheme to personally classify most of the 10,351 stars that would appear in the 1890 *Draper Catalogue*. Though Pickering did acknowledge that "the greater portion of this work, the measurement and classification of all the spectra, and the preparation of the catalog for publication, has been in charge of Mrs. M. Fleming," he reserved sole authorship of the document for himself. Fleming's system comprised a total of seventeen spectral types, beginning with "A stars," which had the most hydrogen, "B stars" with the next largest amount, and so on down to "O stars," with showed virtually no evidence of hydrogen. Her "P" and "Q" stars were categories for objects that did not fit within the A to O pattern. A version of Fleming's spectral types would soon become known as the "Harvard Spectral Classification System," which, with some significant modifications, is still in use today.

Astronomers at the time viewed Fleming's typology as a very powerful tool for progress in stellar astrophysics. Marrying her spectral types to the idea that stars were born hot and cooled off as they aged, they supposed that A stars were the youngest and hottest, O stars the oldest and coolest. One part of this conceptual framework was valid: spectral types are associated with temperature. But that's where the accuracy ended. A stars are not necessarily young and they are not the hottest stars. In fact O stars are the youngest and hottest.

Fleming's great success and the immediately perceived importance of her work led Pickering to hire the rest of his team of women computers and to place them under Fleming's supervision. Henry Draper's niece Antonia Maury, an 1887 Vassar graduate who studied astronomy under the tutelage of Maria Mitchell, Vassar's first professor of astronomy, was given the task of classifying an additional several hundred stars, those for which Pickering's spectra were of the very highest quality. In her work, published in 1897, she rearranged the spectral sequence so that it started with O stars, then B stars, then A stars, followed by the rest of the spectral types, in alphabetical order. This new sequence, OBACDE . . . was still believed to be an aging and cooling sequence (which it was not); but it was getting closer to being a correct temperature sequence. Maury also added four subcategories: a, b, c, and ac (or not-c). Notably, the a and b subtypes had broad hydrogen lines (i.e., the dark spectral lines were fat) while the c and ac subtypes had narrow hydrogen lines.

Annie Jump Cannon, who hired onto Fleming's crew in 1896, would eventually classify over 225,000 stars for the *Henry Draper Catalogue* that would be published in nine volumes over a seven-year period, from 1918 to 1924. But already in 1901, as she was completing her first major effort of classifying over 1,000 stars, Cannon had decided that Fleming and Maury's classifications were inadequate and redundant.

Figure 8.4. Left, Annie Jump Cannon in her Oxford robe in 1925. Right, Cannon at work classifying spectra. Images courtesy of Harvard College Observatory photo collection.

She reordered the categories and eliminated some of the overlapping designations, settling on a classification system that like Maury's began with the O, B, and A stars but continued with the F, G, K and M stars. At Pickering's urging, she dispensed with any further effort to classify stars using Maury's subtypes and also with all other spectral types except for the P (planetary nebulae) and Q (peculiar) stars. Within her revised spectral sequence, Cannon also recognized fractional gradations (O2, O5, O8, B0, B2, B5, etc.), accurate to approximately a quarter of a spectral class.

Thanks to Annie Jump Cannon, at the beginning of the twentieth century astronomers had a one-dimensional spectral typing system that accurately ordered stars from hottest to coolest and (so they thought) from youngest to oldest. The "hottest to coolest" part was right; the "youngest to oldest" was not. But the spectral sequence, now correctly arranged by temperature, was a still great start toward our modern understanding of stars. If an astronomer could identify a star from its spectrum as an O type, he now knew immediately that it was among the hottest of all stars.

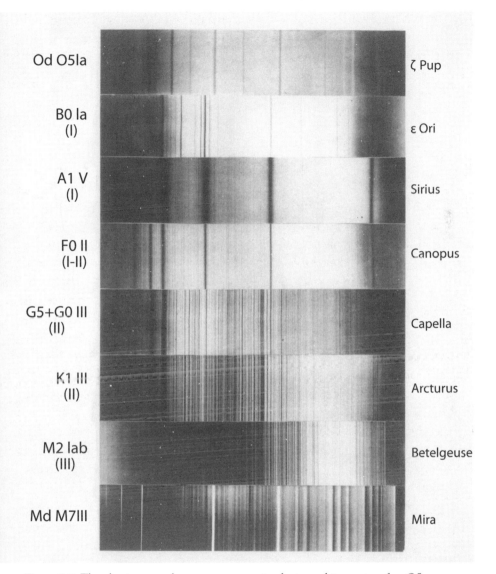

Figure 8.5. The classic spectral sequence as seen in photographic spectra of an O5 star (Zeta Puppis), a B0 star (Epsilon Orionis), an A1 Star (Sirius), an F0 star (Canopus), a G2 star (Capella), a K1 star (Arcturus), an M2 star (Betelgeuse), and an M7 star (Mira). The strong (dark) lines in these negative images for Sirius are due to hydrogen. The signature of hydrogen is weaker going both upwards and downwards in the spectral sequence. The strong lines in the O5 star are due to ionized helium, in the B0 star to neutral helium. The strong lines on the left side of the F, G, and K spectra signal calcium, and the absorption bands in the M7 spectrum indicate the molecule titanium oxide. Image courtesy of Jim Kaler.

What was still missing was a second piece of information by which to classify the stars. That second dimension was supplied in 1905 by the great Danish astronomer Ejnar Hertzsprung, using in part Maury's spectral subtypes, and independently in 1910 by one of the giants of American astronomy, Henry Norris Russell of the Princeton University Observatory.

Giants and Dwarfs

Some stars move relative to other stars, literally changing their locations in the sky in relation to the apparently fixed positions of the overwhelming majority of stars. Astronomers call such movement the star's *proper motion*. Unlike parallax, in which a star returns to the same position in the sky relative to the other stars annually, the proper motion of a star is a steady movement in one direction in the sky, as it is viewed in relation to the apparently fixed positions of most other stars. Why does this happen? Two reasons: first, all stars move through space, but some move faster than others, just as some drivers on the Interstate keep the pedal closer to the metal than do others; second, some stars appear to change positions more rapidly simply because they are closer to the observer. The second reason is by far the more common: in most cases, high proper motion stars are nearby. In a statistical sense, therefore, astronomers can use high proper motion as a proxy for parallax. That is exactly how Hertzsprung made his dramatic breakthrough.

He was able to separate stars of the same color, specifically the yellow (G), orange (K), and red (M) stars, into two groups, one of distant stars, the other of nearby stars. He discovered that the distant stars, namely those stars with either small parallaxes or without measurable proper motions or both, were all intrinsically high luminosity stars. In contrast, the more nearby stars, those with either large parallaxes or high proper motions or both, had much lower luminosities. He found that the yellow stars had the highest average proper motion, and therefore concluded that these stars were closest to the Sun; the faint red stars on average also had large proper motions and therefore were quite

Figure 8.6. Left, Ejnar Hertzsprung. Right, Henry Norris Russell. Images courtesy of Yale University Library (left) and the *AIP Emilio Segre Visual Archives, W. F. Meggers Collection.*

near the Sun. At the other end of the measuring stick were the bright red stars, which had, on average, zero proper motion, making them the most distant group of stars.

This last result was a surprise. How could the brightest stars be the most distant stars? Shouldn't the closest be the brightest? All other things being equal (e.g., the intrinsic brightness of the stars and their physical sizes), in the absence of any interstellar material that would affect their brightnesses, the inverse square law for light demands that more distant stars should be fainter than closer stars. Therefore, if the bright red stars are much farther away than the faint red stars, then the apparently bright red stars must be intrinsically enormously more luminous than the apparently faint red stars. If the color of the star (or the strength of the hydrogen lines in the spectrum) is an indicator of temperature, then the bright red stars and faint red stars have similar temperatures. If two stars are identical in temperature but one is much more luminous than the other, then according to the Stefan-Boltzmann law the more luminous star must be much bigger than the less luminous star.

Hertzsprung had discovered that some stars are big, while others are small. Hertzsprung then discovered that Maury's subtypes in the Harvard spectral sequence, or to be more specific the width of the spectral lines in the different subtypes, made it possible to distinguish intrinsically luminous (narrow-line) stars from the intrinsically faint (broad-line) stars. The spectra thus became a powerful tool for exploring the astrophysics of stars: a red star with narrow spectral lines (Maury's c-type) was an extremely luminous, cool, giant star; a red star with thick spectral lines was a faint, cool, dwarf star.

In 1905 Hertzsprung published his seminal paper "Zur Strahlung der Sterne" [The radiation of stars] in the obscure *Zeitschrift für wissenschaftliche Photographie* [Journal of Scientific Photography]. Over the years 1906–1908, Hertzsprung wrote several letters to Pickering, urging him to reinstate Maury's subtypes in the Harvard spectral sequence that Pickering had had Annie Jump Cannon remove, because they were so enormously useful for distinguishing between big stars and little stars. He failed to persuade Pickering.

In 1910, Henry Norris Russell completed work on a project begun in 1902 in which he made parallax measurements of fifty-two stars, nearly doubling the number with measured parallaxes and distances. With these figures in hand, Russell immediately knew the absolute magnitudes of these stars, and, with their spectra provided to him by Pickering, Russell independently came to the same conclusions as had Hertzsprung: there are both big and little stars of the same spectral types and with the same temperatures; and the red stars are, on average, fainter than stars of other colors. In private letters to Pickering, Russell began referring to these big and little stars as *giants* and *dwarfs*, terms that he would later credit to Hertzsprung and that have subsequently become standard nomenclature.

In an address entitled " 'Giant' and 'Dwarf' Stars" presented to the Royal Astronomical Society in London on June 13, 1913 (and published in *The Observatory* that year), Russell described his now famous diagram in which "the vertical coordinates give the spectra, and the horizontal the absolute magnitude . . . ." This diagram would first be published, with the axes reversed, in a paper in *Nature* in 1914. In the 1930s, it would become known as the Hertzsprung-Russell diagram

Nᵒˢ· 618-619   THE ASTRONOMICAL JOURNAL.   149

OBSERVED PARALLAXES.

No.	Star	R.A. 1900	Decl. 1900	Mag.	Sp.	P.M.	Parallax	p. e.	Pl.	Cmp Star	Observed p. e.	Absolute Magnitude	Cross Vel'y Km/Sec
1	β Cassiopeiae	h m 0 3.8	° ′ +58 36	2.42	F_8	0.56	+0.082	±0.019	5	9	±0.009	2.1	31
2	Groombr. 34	0 12.7	+43 27	7.73	Ma	2.80	+0.250	±0.016	6	9	±0.011	9.8	51
3	26 Andromedae	0 13.3	+43 15	6.04	A	0.03	−0.026	±0.042	6	9	±0.041	...	...
4	η Cassiopeiae	0 43.0	+57 17	3.64	F_8	1.24	+0.187	±0.019	7	8	±0.021	5.1	30
5	o Ceti	2 14.3	− 3 26	var.	Md	0.24	+0.136	±0.035	7	9	±0.035	2.5 to 10.4	8
6	ρ Persei.	2 58.8	+38 27	var.	Mb	0.17	+0.083	±0.040	7	9	±0.040	3.1 to 3.9	9
7	β Persei	3 1.7	+40 34	var.	B_8	0.01	+0.007	±0.027	7	8	±0.025	...	...
8	Lal. 6888 }	3 40.2	+41 9	8.35	G	1.38	−0.029	±0.033	6	6	±0.033	...	...
9	Lal. 6889 }			8.89			+0.020	±0.034			±0.035	...	...

Figure 8.7. First nine lines of a table from Henry Norris Russell's 1910 paper. showing measured values of parallax for stars (Column 8), which he used with the measured apparent magnitudes of these stars (column 5) to calculate absolute magnitudes (column 13). With these absolute magnitudes, Russell had incontrovertible evidence that some stars of a given spectral type are faint (star 4), while others of the same spectral type (star 1) are bright.

that remains, a century later, the most ubiquitous and important diagram in modern astronomy.

In his 1913 address, Russell noted that "there do not seem to be any faint white stars" (lower left area of the diagram), emphasized that "all the very faint stars . . . are very red" (lower right area of the diagram), pointed out that "all the stars of Classes A and B, especially the latter, are many times brighter than the Sun," (the Sun is a spectral type G star) and emphasized that "there is no doubt at all that there exist many very bright red stars (such as Arcturus, Aldebaran, Antares, &c.) [that are] so bright that we can see them at enormous distances." Russell made note of the absence of red stars of comparable brightness (magnitude +5) to the Sun: those [red] stars "are either much brighter or much fainter." In his 1914 *Nature* paper, he wrote, "there are two great classes of stars, one of great brightness . . . the other of smaller brightness." At this point, Russell graciously gave Hertzsprung credit for this discovery and for coining the "giant" and "dwarf" terminology: "These two classes of stars were first noticed by Hertzsprung, who has applied to them the excellent names of *giant* and *dwarf* stars." Most dramatically, Russell asserted that "one can predict the real brightness of a dwarf star from a knowledge of its spectrum alone." Russell's 1913 address to the Royal Astronomical Society was an epochal moment in

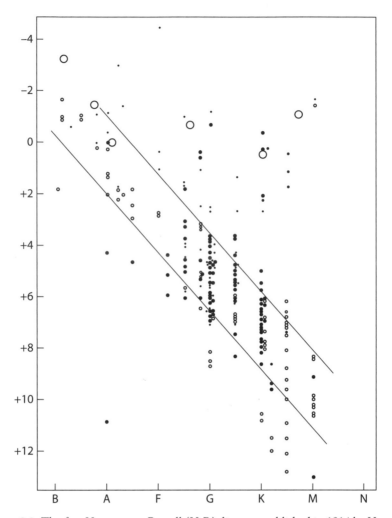

Figure 8.8. The first Hertzsprung-Russell (H-R) diagram, published in 1914 by Henry Norris Russell (*Nature* 93: 252). The horizontal axis indicated spectral type (a proxy for temperature), with the hottest stars at the left, the coolest at the right. The vertical axis shows absolute magnitude, with the brightest stars at the top, the faintest at the bottom. The small solid and open dots represent measurements of individual stars; the large open circles are average measurements for groups of stars (the six open circles include data for 120 stars). The diagram shows the absence of faint white stars (lower left) and shows that the faintest stars are red (lower right). Red stars can however be either faint or extremely bright (top right). Russell noted that although there is some scatter both above and below, most stars cluster between the two diagonal lines in the region that has come to be known as the *main sequence*.

the history of astronomy. It was now possible to securely determine the absolute magnitude—the real brightness—of a star and thereby the distance to the star from its stellar spectrum alone.

Maury's spectral subtypes could, as Hertzsprung had shown, be used to distinguish between giants and dwarfs. For the dwarf stars, astronomers could easily measure apparent magnitudes. To determine distances to these stars, they had up to now needed to measure their parallax angles, measurements that were exceedingly difficult, time consuming, and prone to large errors. Henry Norris Russell now asserted: we do not need parallaxes. We need spectra. From the spectra and the apparent magnitudes, we can determine distances. For Russell, the correlation between the spectral types of dwarf stars and their intrinsic brightnesses was empirical. A firm understanding of the physics of stars would begin to emerge with Hans Bethe's landmark paper "Energy Production in Stars," which was published a quarter of a century later, in 1939. Stellar astrophysics would rest on even more solid ground after World War II, when physicists who had worked in military research and who understood nuclear fusion were able to apply that knowledge to stars. All of modern astronomy, however, rests on the strength of Russell's assertion that spectral types of dwarf stars and their luminosities are correlated. It is crucial that we understand why this statement is true.

· · · CHAPTER 9 · · ·

Reading a Hertzsprung-Russell (H-R) Diagram

The line spectrum of the companion is identical to that of *Sirius* in all respects . . . indicating that the companion of *Sirius* has a color index not appreciably different from that of the principal star.

—Walter Adams, in "The Spectrum of the Companion of Sirius," *Publications of the Astronomical Society of the Pacific* (1915)

The Hertzsprung-Russell diagram is so fundamental to virtually everything astronomers have learned about stars, galaxies, and the universe that it's worth taking some extra time to make sure we understand exactly how powerful a tool it is. Remember that Henry Norris Russell's first diagram, published in 1914, plotted the absolute magnitude of a star on its vertical axis and its Harvard spectral type on the horizontal axis. A century later, the basic approach for how astronomers make such a diagram of luminosity versus temperature has not changed much; however, astronomers have found several other observational measurements that they often use instead of the absolute magnitude (e.g., luminosity in comparison to the Sun) or instead of the Harvard spectral type (stellar color; see Figure 9.1).

To put any one star on an H-R diagram, an astronomer needs to determine values for the temperature and absolute brightness of that star. The spectral type of a star, which is a direct measurement of the temperature of the surface of the star, can be determined in a straightforward manner, provided one can collect adequate light to obtain a spectrum for the star. For a faint star, an astronomer might need to use a bigger telescope or make a longer time exposure on a photographic plate (or, in the twenty-first century, on a CCD chip) to

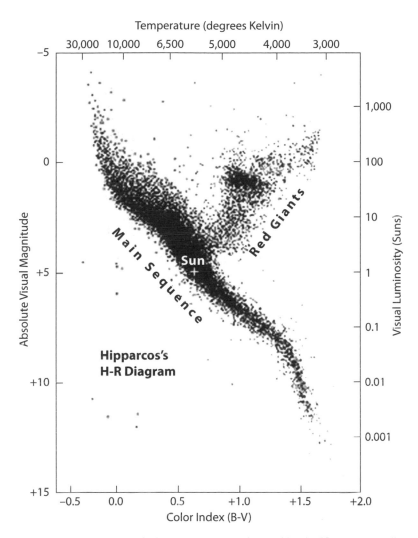

Figure 9.1. H-R diagram made from 20,853 stars observed by the Hipparcos satellite. The temperature (equivalent to spectral type) is marked along the horizontal axis at the top of the plot. The surface temperature of a star can also be estimated fairly accurately by measuring the intensity (or magnitude) of light from the star in two different broadband filters (for example, U for ultraviolet, B for blue, V for visible, R for red) and calculating the difference between these two magnitudes. Astronomers call this difference a *color* or *color index* (e.g., the B-V color; say "B minus V color"). A star with a large B-V color is very red and cool; one with a negative B-V color is very blue and hot. This color (marked along the bottom of the plot) is unique for each temperature across the broad range of temperatures that characterize stars. Thus, the horizontal axis of an H-R diagram could be plotted as a function of spectral type, temperature, or color. The brightnesses of the stars are plotted along the vertical axis, either in terms of the absolute magnitude (left) or in comparison with the luminosity of the Sun (right). Image courtesy of Michael Perryman.

collect enough light, but no other knowledge about the star is needed to determine its location along the horizontal axis. The brightness of a star, however, is more difficult to determine, as one must know both its apparent magnitude, which is easy to measure, and the distance to the star, which as we have seen is very difficult to measure; but given both an apparent magnitude and a parallax angle, the absolute magnitude can be calculated directly from the inverse square law for light.

The most important feature in an H-R diagram is the band of stars running from the lower right (faint, cool, red stars) to the upper left (bright, hot, blue-white stars). This strip has come to be called the *main sequence*. And in the language of astronomers, all main-sequence stars are referred to as dwarf stars, even though the hot, blue ones are much bigger than the cool, red ones. At the upper right, we find a less well-populated region of bright, cool, red stars; at the lower left, we find a sparsely populated region of faint, hot, blue-white stars.

Comparing Up with Down on the H-R diagram

To get a better feeling for the H-R diagram, let's compare two stars of the same spectral type. For both, we have measured the apparent magnitude and the parallax angle, which gives us the stars' absolute magnitudes or brightnesses. In terms of absolute brightness, one star (Star B) is much fainter than the Sun while the other (Star A) is more luminous than the Sun. Star A is therefore much more luminous than Star B. We do not know the temperature of either star, but because they are of the same spectral type we do know that they have identical temperatures and therefore would emit the same amount of light if they were the same size. It follows that the only reason Star A is brighter than Star B is because it has a larger emitting surface area than Star B. Star A is a giant star and Star B is a dwarf star. If both stars were M stars and therefore appeared red, we would call Star A a red giant while we would refer to Star B as a red dwarf.

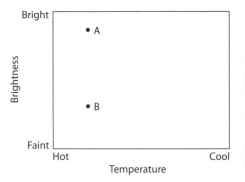

Figure 9.2. A comparison of two stars, A and B, on an H-R diagram. Stars A and B have the same temperature, but Star A is much brighter than Star B. Therefore, Star A must be much larger (with more surface area from which to emit light) than Star B.

Comparing Left with Right across the H-R Diagram

Now, let's compare two stars of the same absolute magnitude but of different spectral types. Star A is spectral type B0 while Star C is spectral type M0. From a post-1913 understanding of why certain spectral lines appear at certain temperatures, theorists have calibrated absolute temperatures for all spectral types. As a result, we know that Star A has a high a surface temperature (30,000 K) while Star C has a much lower surface temperature (4,000 K); Star A therefore emits much more energy per square meter of surface area than does Star C. Since these two stars have the same absolute magnitude, they give off the same total amount of light; however, since Star A is hotter than Star C, Star A emits light much more efficiently than Star C from every square meter of its surface. We can therefore conclude that Star A must be significantly smaller in surface area than Star C. In other words, when comparing two stars with the same intrinsic brightness, the red star is much bigger than the blue star. If both stars were intrinsically in the upper range of brightness, Star A would be a blue giant while Star C would be a red giant.

How Small Can a Star Be?

Alvan Graham Clark was an astronomer, telescope-maker, and the son and father of telescope-makers in the famous Alvan Clark & Sons

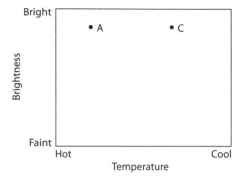

Figure 9.3. A comparison of two stars, A and C, on an H-R diagram. Stars A and C emit the same total amount of light, but Star A is much hotter than Star C and consequently emits much more light from every square meter of its surface than Star C. Therefore, Star C must be larger (with more surface area from which to emit light) than Star A in order to compensate for emitting light less effectively.

Company. Their most famous telescopes, both of which used large lenses (modern telescopes use mirrors) to collect and focus starlight, are the 36-inch (the diameter of the glass lens) Lick Observatory refractor in California, completed in 1887, and the 40-inch Yerkes Observatory refractor in Wisconsin, completed in 1897. When completed, each was the largest telescope in the world at that time. In 1862, Clark discovered a faint companion to Sirius whose existence Friedrich Bessel had predicted two decades earlier, based on his observations of the irregular motion of Sirius. Since Sirius and its faint companion, now called Sirius B, are members of a binary star system, they are at very nearly identical distances (2.64 parsecs) from the Earth and Sun; therefore, any difference in their apparent brightnesses must indicate the same difference in their intrinsic brightnesses. Furthermore, this discrepancy must be due to a difference between either their temperatures or their sizes or both.

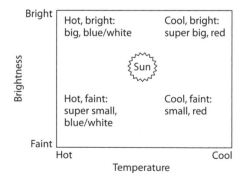

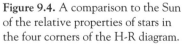

Figure 9.4. A comparison to the Sun of the relative properties of stars in the four corners of the H-R diagram.

Figure 9.5. Sirius (center) and Sirius B (toward lower left corner). Image courtesy of the Space Telescope Science Institute.

Sirius is about twenty-three times more luminous than the Sun at the wavelengths of visible light. The temperatures of Sirius and the Sun are about 9,900 K and 5,780 K, respectively. With this information and the Stefan-Boltzmann law (Chapter Seven), we can directly determine the diameter of Sirius: 2,300,000 kilometers, which is 60 percent greater than the diameter of the Sun. Now, with knowledge of the absolute size of Sirius, we can use the relative brightnesses and temperatures of Sirius and Sirius B to determine the absolute size of Sirius B.

Including light emitted by both stars at all wavelengths, we know that Sirius is 780 times brighter than Sirius B. In 1915, Walter Adams determined that the temperatures of Sirius and Sirius B are nearly identical; this measurement led directly to the realization that the size of Sirius B was comparable to or even smaller than that of the Earth. The most accurate modern measurements reveal the surface temperature of Sirius B to be about twice as hot as that of Sirius, at about 25,000 K. This information, along with the Stefan-Boltzmann law and our knowledge of the actual physical size of Sirius allows us to calculate the size of Sirius B. The answer: Sirius B has a diameter of about 12,800 kilometers, which is virtually identical to the diameter of the Earth

(12,756 km at the equator, while being more than four times hotter but about thirty-four times less luminous than the Sun. These properties are typical of the stars we now call white dwarfs.

This dramatic conclusion, that a kind of star exists that is the same size or perhaps even smaller than the Earth yet hotter than the Sun, comes directly from a firm understanding of the meaning of spectral types (they are determined by the surface temperature of the star) and the assertion that the spectral type of a star determines its absolute luminosity.

Red giants, blue giants, red dwarfs, white dwarfs: direct proof that all these exotic objects exist comes directly from the H-R diagram, the Stefan-Boltzmann law, and the measurements astronomers make to place stars, as data points, on the diagram. In 1914, astronomers had the knowledge that enabled them to "read" the H-R diagram and discover the sizes of stars. What they did not have yet was a clear understanding as to why stars of different sizes, temperatures, and luminosities appeared in their respective places on this diagram.

· · · CHAPTER 10 · · ·

Mass

It appears therefore that, as a matter of statistical averages, the masses of the stars depend upon their absolute magnitudes, rather than upon their spectral types. The brighter stars are the more massive.

—Henry Norris Russell, in "On the Mass of the Stars," *Popular Astronomy* (1917)

Why are some stars hotter than others, and why are the hotter stars more luminous than most of the cooler stars? At the beginning of the twentieth century, most astronomers answered this question by saying that stars are born hot and luminous. According to this explanation, at their birth all stars appear in the top, left corner of the H-R diagram. As they age, they lose energy and cool off, thereby moving downwards and to the right. But as tidy as this answer may have sounded, it failed to explain the simultaneous existence of both red dwarfs (stars moving downward and to the right, as would be expected, as they age) and red giants (stars moving upwards and to the right with age).

Measurements made in the first decades of the century suggested a better answer, one that was proposed as early as 1914 by Henry Norris Russell. Simply put, the mass of a star determines both its temperature and luminosity, and thus its location on the H-R diagram: stars born with more mass are hotter and more luminous than stars born with less mass. But before Russell could reach this conclusion, astronomers had to develop the tools for measuring the masses of stars.

Redshifts and Blueshifts

In 1842, Austrian mathematician and physicist Christian Doppler published a monograph in which he predicted that the frequency (or wavelength) of a sound wave would be affected by the motion of the source of the sound toward or away from the observer of the sound. Soon thereafter, in 1848, French physicist Louis Fizeau made a similar prediction about light, suggesting that the wavelengths (or colors) at which starlight arrives in our telescopes will be affected by the motion of the star toward or away from the Earth. These changes in the wavelengths of light or sound caused by the motion of the light (or sound) source toward or away from the observer are known as Doppler shifts.

Here's an example of a Doppler shift: if the velocities of a star and of the Earth are such that the distance between them is increasing (we can think in terms of the Earth moving away from the star or the star moving away from the Earth; all that matters is their relative motion), the wavelength of light from that star *as observed by an astronomer on Earth* will be larger than the wavelength at which the light *was emitted by the star*. If we think about yellow light (in the middle of the visible spectrum) emitted by such a star, the yellow light would appear shifted toward longer (redder) wavelengths as seen by an Earthly observer. Thus, we call the change in the wavelength of light a *redshift* if the distance between the Earth and the star is increasing. On the other hand, if the velocities of a star and the Earth are such that the distance between them is decreasing, the wavelength of light from that star that is observed by an astronomer on Earth will be smaller than the wavelength at which the light was emitted by the star. In this case, the yel-

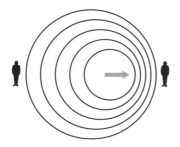

Figure 10.1. Illustration of Doppler shift, with light source moving to the right.

low light emitted by the star would appear shifted toward the blue, producing what we call a *blueshift*. The terms "blueshift" and "redshift" are now used to designate any wavelength shift of light toward shorter or longer wavelengths, respectively, that is due to the changing distance between the source of the wave and the observer, whether we are referring to X-rays, visible light, or radio waves.

In the most familiar situations in which we encounter redshifts and blueshifts, the shift is due to the distance between the light source and the observer changing because the objects are moving away from or toward each other through space (astronomers refer to the motion of an astrophysical object toward or away from the Earth, or more generally to the increase or decrease in separation over time between the Earth and the object, as the *radial velocity* of that object). In these situations, we call the redshifts and blueshifts *Doppler shifts*. We shall see later that if space itself is expanding, the distance between the light source and the observer will also increase, producing a redshift that we term a *cosmological redshift* rather than a "Doppler shift." A third physical mechanism that results in a redshift is gravity; when light is emitted by a very massive object, the gravity of that object saps energy from the emitted photon. Because the speed of the emitted photon cannot change, the loss of energy from the photon manifests as a redshift, known as a *gravitational redshift*.

In practice, astronomers do not look for yellow light that appears shifted to the red or blue. Instead, they look for specific emission and absorption lines, for example the set of emission lines produced by hydrogen gas at temperatures typical of stars as measured in the laboratory, in the spectra of the objects they study. Laboratory experiments in which the hydrogen gas is at rest (not moving toward or away from the Earth) provide precise measurements of the wavelengths at which hydrogen atoms normally emit light (their "rest" wavelengths). From spectra of stars, astronomers can identify an entire suite of hydrogen lines, measure the wavelengths at which those lines are detected and compare the measured wavelengths to the rest wavelengths. The greater the difference between the measured and rest wavelengths, the greater the radial velocity of the astrophysical object. Since the spectral lines astronomers observe are the fingerprints of elements and

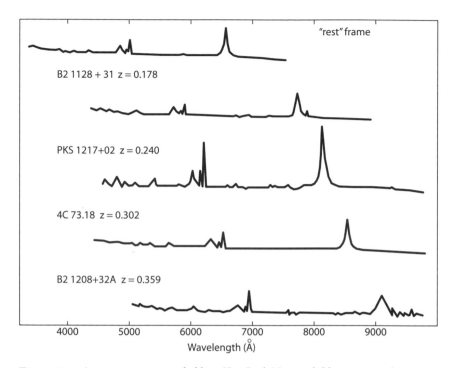

Figure 10.2. A spectrogram recorded by a Kitt Peak National Observatory telescope, in which the top spectrum (obtained for a hydrogen light source at rest in the laboratory) has peaks for three hydrogen lines in the blue (4,340 angstroms), green (4,861 angstroms), and red (6,563 angstroms). The bottom four spectra are from distant quasars at progressively greater distances and show the hydrogen lines shifted ever greater distances toward the right (red) end of the spectrum. Image by M. Corbin.

molecules at particular temperatures and densities, the rest wavelengths for a vast array of elements and molecules that might appear in astrophysical spectra have been measured in laboratory experiments and are available for comparison with telescopic data.

Redshift or blueshift is of great importance for measuring the radial velocity of an object. This is because the amount of the shift does not depend on any other properties specific to that object (e.g., temperature, mass, diameter, composition, type of object) but on its radial velocity alone. Since radial velocity is measured directly from a spectrum, and since this measurement does not require knowledge of the distance to or the absolute magnitude of an object, radial velocities can be measured for all objects that are bright enough that we can obtain spectra for them. Even if an object is intrinsically faint or very faint because it is exceedingly distant, we can still determine its radial velocity, provided we are willing and able to build a big enough telescope or take sufficiently long time exposures (or both) in order to collect enough light to make the measurement.

Using Radial Velocity Measurements to Measure Stellar Masses

The Earth orbits the Sun at a nearly constant speed of about 30 kilometers per second. If we were observing the Earth-Sun system from far away, perhaps from another star system in our galaxy, and from a vantage point in the ecliptic plane (the Earth's orbital plane around the Sun), we might discover that the Earth was moving away from us (the light from the Earth would be redshifted) at a speed of 30 kilometers per second. In a matter of days, however, we would observe the speed of the Earth away from us beginning to slow down. After three months, the speed we measure of the Earth away from us would have dropped to 0 kilometers per second, as the Earth would now be moving across our line of sight rather than away from us. Then, Earth's measured speed would become negative (the light from the Earth would be blueshifted), reaching a maximum speed toward us of thirty kilometers per second after three more months. Over the next three months, Earth's measured speed would increase back to 0 kilometers per second and then it would continue to increase for three more months until it reached 30 kilometers per second again. If we collected spectra of the Earth-Sun system year after year, we would see this cycle (redshifted to blueshifted to redshifted) repeat every 365.25 days.

In this example, the radial velocity of the Earth as measured by distant observers would change from positive to zero to negative to zero. The Earth, however, would never actually be changing speed; it would simply be changing its direction relative to the viewer. If, on the other hand, we were observing the Earth-Sun system from above the north pole of the Sun, we would be able to watch the Earth orbit the Sun;

but the radial velocity of the Earth as measured by us would remain constantly at zero, since the Earth would never move toward or away from us.

We need now to look again at Kepler's third law (see Chapter Five), which specifically says that the square of the orbital period (the number of years it takes for a planet to orbit the Sun, multiplied by itself) is equal to the cube of the semi-major axis (the semi-major axis multiplied by itself and then multiplied by itself again). Kepler's third law describes how planets orbit the Sun under the control of the force of gravity. Three quarters of a century after Kepler wrote down this law, Isaac Newton showed that it could be derived mathematically directly from Newton's law of gravity, which explicitly includes the masses (actually, the sum of the two masses) of the two orbiting objects (e.g., the Sun and one planet or one star orbiting a second star) and a constant of proportionality known as Newton's gravitational constant. (Because Newton's version of Kepler's third law includes the masses of the orbiting objects explicitly, *it is valid for any two objects in orbit around each other*, whereas Kepler's third law as written down by Kepler only works for objects orbiting the Sun.)

The power of Newton's version of Kepler's third law is that if we already know the size of the orbit and the orbital period for two binary stars (or indeed any two objects) orbiting around each other, we can use those measured parameters to calculate the only other unknown: the sum of the masses of the two stars. This is an amazing piece of physics. Simply by measuring how long two stars take to orbit each other and by measuring how far away the binary system is from us (from the measured parallax for the stars) and the angular separation of the two stars (which we can convert to a physical distance if we know the distance to the system), we can measure the total amount of mass of the two stars. Unfortunately, this measurement does not, by itself, give us the masses of either one of the two individual stars, just the sum of their masses. In very special cases, however, we can determine the individual masses. Those special cases are eclipsing binary star systems for which we can measure radial velocities.

Beginning with the work of William Herschel, astronomers struggled tirelessly for most of the nineteenth century to discover and catalog binary star systems. To calculate masses of stars, though, not just any old binary system would do. Astronomers had to know the distance to the binary (the distance allows astronomers to convert the angular separation between the two stars into a true physical distance) and they needed additional information about the orbits, preferably knowledge about the orientation of the stellar orbits (eclipsing or non-eclipsing) and the radial velocities of the stars. A few such systems were known by the early twentieth century.

Stellar Masses, the H-R Diagram, and the Mass-Luminosity Relationship

In 1910, American astronomer Robert Grant Aitken of Lick Observatory, the most eminent double-star expert of the early twentieth century, compiled a list of those binary systems for which astronomers claimed to know the masses of both stars. His list included only nine, for a total of eighteen stars. The most massive star on his list was 4.04 solar masses (M_{sun}) and the least massive a minuscule 0.22 M_{sun}. Within a few years, knowledge of the masses of stars in eclipsing binary systems had increased dramatically as a result of work by Harlow Shapley for his doctoral dissertation work, done at Princeton under the supervision of Henry Norris Russell, and follow-up work that Shapley published in 1913 and 1914. Thanks to the contributions of Shapley and Russell, stellar mass calculations became reasonably accurate. For example, for the system Mu Herculis, Shapley determined that the stars had masses of 7.66 times the mass of the Sun (7.66 M_{sun}) and 2.93 M_{sun}. In 2004, those masses were redetermined to be 7.85 M_{sun} and 2.85 M_{sun}.

Henry Norris Russell compiled his own list of stellar mass measurements in 1917. His list by now included 113 binary star systems, a figure large enough to allow him to draw a number of brash conclusions, including the following: the average mass of the most luminous B stars is about 17 M_{sun}; the average mass of less luminous B stars and A stars is about 5 M_{sun}; the average mass of F dwarf stars is about 3.5 M_{sun};

TABLE 10.1.
Stellar Masses: Russell Breaks the Code

Spectral Type	Average Stellar Mass
B	17 M_{sun}
A	5 M_{sun}
F	3.5 M_{sun}
K and M	< 1 M_{sun}

and the average mass of K and M dwarf stars is less than 1 M_{sun}. "It appears therefore," he wrote, "that, as a matter of statistical averages, the masses of the [dwarf] stars depend upon their absolute magnitudes. . . . The brighter stars are the more massive." Barely three years after the first H-R diagram appeared in its nearly modern form, Russell had broken the code for the dwarf stars: the mass of a star determines both its luminosity and its surface temperature. More simply put: *the location of a dwarf star on the H-R diagram is controlled by a single parameter: its mass*. This was the moment when the H-R diagram became the single most important tool for twentieth-century astronomy. It is, in fact, the foundation upon which astronomers have measured the age of the universe.

With historical hindsight however, our statement about dwarf stars—"mass determines location on the H-R diagram"—turns out to be only about 99 percent correct. Small changes in the elemental composition of a star can affect minor changes in its temperature and luminosity. But at the most fundamental level, *for dwarf stars*, the two measurable parameters, temperature and luminosity, depend almost entirely on mass. By 1924, Arthur Eddington had developed from theoretical principles a *mass-luminosity relationship* that fit the known data for main-sequence stars. According to this mass-luminosity relationship, the luminosity of a main-sequence star depends on the mass of the star raised to the 3.5 power (cubed and then multiplied again by the square root of the number). This relationship says that a star with twice the mass of the Sun is not twice as luminous as the Sun. Instead, we find that it is more than eleven times more luminous than the Sun; and a star with ten times the mass of the Sun is about 3,000 times more luminous than the Sun.

Understanding the H-R Diagram:
It's All about Mass

If the mass of a main-sequence star determines both its surface temperature (spectral type) and luminosity and consequently the location of this star on an H-R diagram, then we can determine the location of a main-sequence star on an H-R diagram if we know only its spectral type. Why? Every main-sequence star of a given mass, no matter whether the distance to that star is 42 parsecs or 42 kiloparsecs, will have the same surface temperature and spectral type. Every main-sequence star of that spectral type will have the same luminosity. Therefore, if I measure the spectral type *and can say with confidence that the star is a dwarf and not a giant* (Antonia Maury's spectral subtypes, which were discarded from the Harvard spectral catalog when Annie Jump Cannon revised it, had enabled Hertzsprung to quickly distinguish between dwarf and giant stars), then I know the luminosity of that star.

The importance of this discovery can scarcely be exaggerated. Astronomers can very easily measure the apparent magnitude of a star. Recall, however, that in order for us to measure the absolute magnitude for that star, we also need to know its distance, which we calculate by measuring its parallax. Recall, also, that measuring parallax is exceedingly difficult, time consuming, and in 1910 was only possible for stars within a few tens of parsecs of the Sun. (In 2010, that limit has been extended, but only to a few hundred parsecs.) Spectral types, however, are easy to measure. An astronomer does not need to know the distance to the star; he needs only to collect enough light to produce a spectrum. The spectrum directly gives a spectral type. The spectral type directly yields the luminosity of the star, i.e., its absolute magnitude, if we know that the star is on the main sequence. The absolute and apparent magnitudes together yield the distance to the star. Let's repeat this, because it is of such fundamental importance to all of modern astrophysics:

we can calculate the distance to any main-sequence (dwarf) star by measuring its spectral type and apparent magnitude.

This remarkable astrophysical tool rests on one foundational idea:

> The position of a main-sequence (dwarf) star on an H-R diagram depends almost entirely on the mass of the star.

No matter where stars are in the universe, provided the laws of physics (e.g., gravity, quantum mechanics, the strong nuclear force) are the same everywhere, stars will behave in the same way, controlled almost entirely by how much mass they contain.

But as wonderfully helpful as this discovery proves, for any single star we still have a significant question to answer. How do we know that a red star is on the main sequence (a dwarf star) and not above it (a red giant) or that a hot, white star is on (a dwarf) and not below (a white dwarf) the main sequence? The answer: star clusters.

· · · CHAPTER 11 · · ·

Star Clusters

For the Galaxy is nothing else than a congeries of innumerable stars distributed in clusters.

—Galileo Galilei, in *Sidereus Nuncius* [Sidereal Messenger] (1610)

Take the most casual glance at the nighttime sky and you will quickly notice that the stars are not distributed uniformly from horizon to horizon. Almost 3,000 years ago Homer, in his *Iliad*, and Hesiod, in his *Works and Days*, had already mentioned clusters of stars in their writings: the Pleiades and the Hyades. Any careful observer, equipped with even the smallest telescope, will immediately perceive, as did Galileo, that some objects that appear to the naked eye as stars are actually clusters of stars. Galileo identified a star cluster he called the Nebula of Orion (or Orion's Head), with at least twenty-one stars, as well as the Nebula of Praesepe, with more than forty stars. He also noted that "the stars that have been called "nebulous" by every single astronomer up to this day are swarms of small stars placed exceedingly close together."

Such nebulae apparently were of little interest to seventeenth-century astronomers, who were kept busy enough measuring solar eclipses, discovering the moons (some of them anyway) and the rings around Saturn and the Great Red Spot on Jupiter, mapping the surface of the Moon, and measuring the parallax of Mars and the size of the solar system, among other problems. In fact, "nebulous" stars were something of a nuisance. They could be confused with and therefore hindered efforts to quickly and, with certainty, identify comets, which were objects of great interest in the seventeenth and eighteenth centu-

ries. By the early twentieth century, however, star clusters had become
the keys to the distant universe. With their help astronomers would
unravel the history of the stars and galaxies and discover the age of the
universe itself.

Identifying and Cataloging Star Clusters

Charles Messier, who spent most of his career as Astronomer of the
Navy for the government of France, was more interested in comets
than in nebulae. Comets and nebulae, however, are both nebulous objects, so Messier found it necessary to catalog the positions of nebulae
so as to eliminate the possibility that he and other observers might
mistake them for the more interesting objects that they hoped to (and
would) discover. His work culminated with his "Catalog of Nebulae
and Star Clusters," published in the *Memoirs* of the French Academy of
Sciences in 1774. There Messier listed the positions in the sky for
forty-five such objects. His expanded 1781 edition of this catalog includes 103 objects, all of which are known to modern astronomers by
their Messier catalog numbers. M31 is the Andromeda Galaxy and
M42 is the Orion Nebula, located in the sword of Orion.

A century later, John Herschel published the 1864 *General Catalog*
identifying 5,000 nebulae; and twenty-four years after that, John Louis
Emil Dreyer, director of the Armagh Observatory in Ireland, published
his *New General Catalogue*, a compilation of the locations of the more
than 13,000 nebulae and star clusters that were known at that time.
Star clusters were now poised to burst onto the astronomical scene as
among the most important objects in the heavens.

In 1859, John Herschel set forth a distinction between two distinct
kinds of star clusters, which he called *globular clusters* and *irregular clusters*. Globular clusters are centrally condensed, spherical objects that
contain a virtually uncountable number of stars. The cores of globular
clusters could not be resolved into individual stars in the nineteenth
century, but we now know that the largest globular clusters contain up
to a million stars. Irregular clusters on the other hand contain a more

Figure 11.1. The open cluster NGC 265, located in the Small Magellanic Cloud. Image courtesy of E. Olszewski, University of Arizona, the European Space Agency and NASA.

manageable number of stars—a few dozen or a few hundred—and have no preferred shape. They would later become known as *open clusters*. (In 1930, Robert Trumpler of Lick Observatory suggested the term *galactic clusters* as the complementary term to globular clusters, since open clusters were always found in or very near the plane of the Milky Way. That term was widely used for most of the twentieth century, but now that such clusters can be observed in other galaxies, "open cluster" has again become the preferred name.)

Figure 11.2. The globular cluster M80, located in the Milky Way. Image courtesy of the Hubble Heritage team (AURA/STScI/NASA).

Held Together by Gravity

What is the nature of these star clusters? In 1767, Reverend John Michell of Yorkshire, who was also a geologist and amateur astronomer, calculated that the probability of a chance alignment of stars as close together as those in the Pleiades was only one in 496,000. The existence of such clusters or, as he put it, fluctuations in the spatial density of stars in the heavens, results from either "the original act of the Creator or in consequence of some general law (such perhaps as gravity)."

Michell was right: it is gravity that holds these clusters together. Clusters with large numbers of stars (globular clusters) can hold themselves together for the entire age of the universe, while clusters with fewer stars (open clusters) disperse after only a few tens or hundreds of millions of years.

In 1869, just over a century after Reverend Michell, Richard Proctor wrote in the *Proceedings of the Royal Society of London* that "in parts of the heavens the stars exhibit a well-marked tendency to drift in a definite direction." Proctor had discovered that the stars in a single star cluster share an average motion, and that when this motion is in a direction away from Earth, all the stars appear "to drift" in the same direction. Imagine that you are looking westward while standing on an overpass above a twelve-lane freeway watching cars zoom off toward the horizon. Six cars, one in each of the six westbound lanes, are moving at identical speeds of 100 kilometers per hour while traveling side by side and staying in their own lanes. If you magically transport yourself to another overpass ten kilometers westward, then in six minutes you would see these same six cars pass beneath you, still moving westward in parallel lines at identical speeds. The cars have maintained their positions in their respective lanes; they are neither closer together nor farther apart in actual physical distance. From your perspective atop a single freeway overpass, however, those six parallel lanes appear to converge toward a point on the distant horizon, and so all six cars appear likely to crash when the road meets the horizon. Similarly, if a group of stars are all at a similar distance from the Sun and are all moving in the same general direction away from the Sun at similar speeds, their collective motions will make them appear to converge toward a single point located a great distance away from us. Furthermore, cars (or stars) that are close to us will appear to converge faster than cars (or stars) that are more distant, so this method can be used to estimate physical distances to groups of objects. Astronomers call this the *convergent point phenomenon* and can use the geometry derived from it to determine the average distance to the stars in a cluster.

In 1910 Lewis Boss, who was the director of Dudley Observatory in New York from 1876 until his death in 1912, published a comprehensive catalog of stellar "proper motions" (the apparent vertical and hor-

Figure 11.3. The scattered group of stars in the central part of the image is the open star cluster called the Hyades in the constellation of Taurus. The Hyades can be seen easily by the naked eye as a "V" shape in the sky. The brightest star, Aldeberan, which represents the red eye of Taurus the bull, does not physically belong to the cluster. Image courtesy of Till Credner, AlltheSky.com.

izontal motions of a star across the sky), the *Preliminary General Catalogue of 6188 Stars for the Epoch 1900.* Boss had available to him not only the most accurate proper motion data ever assembled, but also, from Doppler shifts obtained from stellar spectra, the first few measurements of the radial velocities (the component of a star's velocity directly toward or away from the Sun) of three stars in the Hyades Cluster. In 1908 Boss produced a definitive analysis of the three-dimensional motion of these three stars, demonstrating that they have a common radial velocity of 39.9 kilometers per second, a common space velocity (the three-dimensional motion of the stars) of 45.6 kilometers per second, and a clear convergent point that places the cluster at a distance from the Sun of about 38 to 40 parsecs. (This estimate was correct to within about 20 percent; the modern measurement for the distance is about 46 parsecs). The Hyades Cluster stars themselves occupy a volume of space only about 8 to 10 parsecs across and 8 to 10

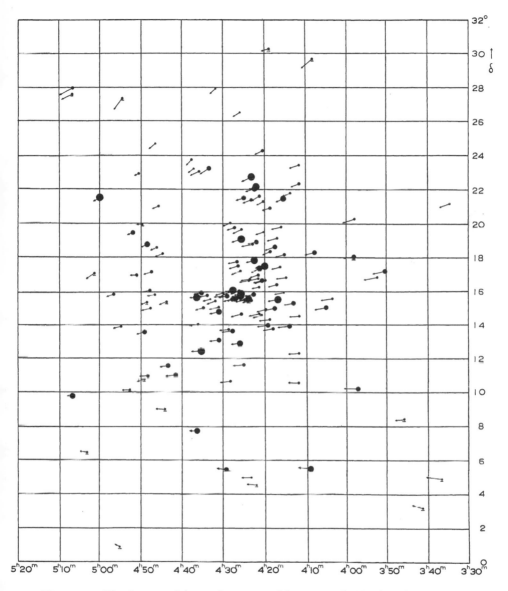

Figure 11.4. The directional (proper) motions of the stars in the Hyades Cluster, depicted with arrows indicating the direction in which each star is moving. The stars appear all to be moving toward a single "convergent" point, located to the left of the left edge of the graph. From H.G. van Bueren, *Bulletin of the Astronomical Institutes of the Netherlands* 11 (May 1952): 385.

parsecs deep, which means that all the many dozens of stars in the cluster are located at about the same distance (46 ± 5 parsecs) from the Sun. In other words, the distances between stars in the Hyades star cluster are much less than the distance from any one of the cluster stars to the Sun. For practical purposes, the statement "all the stars in the Hyades Cluster are at the same distance from the Sun" is quite reasonable. Boss suggested "that the cluster was condensed from a vast nebula that originally had the present velocity and direction of the cluster."

By 1910, a clear understanding of star clusters had emerged. Any single star cluster is a group of stars held together, perhaps only temporarily, by gravity. The stars in the cluster move in a common direction in space, probably because they were born at about the same time from a single, vast, interstellar cloud and therefore share the original motion of that precursor cloud. Since the separations of the stars within a star cluster are much less than the distance from the Sun to any one of the stars in the cluster, we may think of all the stars in a cluster as being at about the same distance from the Sun. For the Hyades, which is quite near the Sun, the spread in distances within the cluster is about 20 percent of the distance to the cluster from the Sun. Assuming other clusters are similar in size to the Hyades but further away from the Sun, the statement that all the stars in a cluster are about the same distance from the Sun becomes more accurate the more distant the cluster.

Main-Sequence Fitting

Our modern understanding of star clusters is built on three basic ideas developed a century ago. First, star clusters are groups of stars (hundreds of stars in the case of open clusters; tens of thousands or even hundreds of thousands of stars for globular clusters) held together by gravity (for tens of millions of years for open clusters; for tens of billions of years for globular clusters). Second, the stars in each cluster formed at about the same time (that is, across a span of at most a few million years) out of a single interstellar cloud of gas and dust that fragmented, with each fragment condensing under the force of gravity to give birth to one star or to a binary or multiple-star system. Third, due to the randomness of the physical processes that fragment interstellar clouds, the stars that form have different masses; the least massive are a few to 10 percent of the mass of the Sun and are not very luminous (as faint as about 1 percent the luminosity of the Sun) while the most massive stars can be as much as fifty times more massive than the Sun and are extremely luminous (as much as 100,000 times more luminous than the Sun). Provided that a cluster contains at least a few dozen stars, it will include stars spanning a wide range of masses and therefore of luminosities and spectral types.

A century ago astronomers could measure the distances to the stars in the Hyades Cluster using the convergent point method. With these distances and the observed apparent magnitudes and spectral types for these stars, constructing an H-R diagram for the Hyades Cluster was a straightforward exercise. The resulting diagram turns out to be very similar to an H-R diagram for all of the stars that happen to be near enough to the Sun for us to have direct measurements of their parallax-based distances and absolute magnitudes; however, those nearby "field" stars do not comprise a star cluster. While taken together they do reveal a main sequence, including some giant stars and some white dwarfs, the field stars did not form together and at the same time out of a single interstellar cloud. And yet, the Hyades Cluster, all by itself, shows the same features. Not surprisingly, the absolute magnitude of the G stars in the Hyades is the same as that of the G field stars, and the absolute magnitude of the F stars in the Hyades is the same as that of the F type field stars. In fact, the entire main sequence of the Hyades matches the main sequence found for the field stars. Of course, we know this must be the case since we know that the luminosities and temperatures of main-sequence (i.e., dwarf) stars are determined almost entirely by their masses; therefore, every main-sequence G star in every star cluster should have the same luminosity and temperature as every other main-sequence G star in every other cluster—and the same should be true of every main-sequence O, B, A, F, K, and M star.

So then, what can we learn about the stars in a cluster, even if they are so far from the Sun that making parallax measurements to determine their distances and absolute magnitudes is impossible? We know

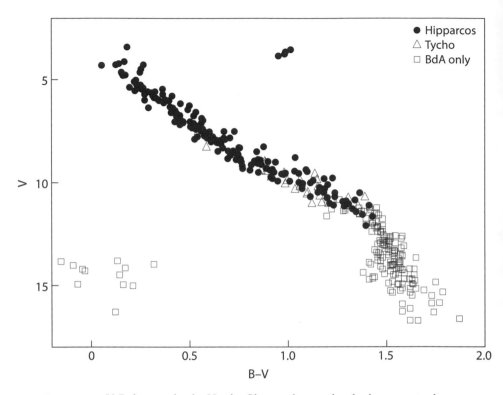

Figure 11.5. H-R diagram for the Hyades Cluster, plotting the absolute magnitude in the V (visual) band versus the B-V color. (Versions of H-R diagrams that use B-V color instead of spectral type are known as color-magnitude diagrams. Color-magnitude diagrams plot actual observed quantities while H-R diagrams plot quantities [temperatures] calculated from the observed colors or spectra for the stars.) The diagram shows the main sequence for the Hyades, along with some red giants (top right) and white dwarfs (bottom left). From Perryman et al., *Astronomy & Astrophysics* (1998) 331, 90. Reproduced with permission © ESO.

that without parallax measurements, we have only spectral types and apparent magnitudes, and we had decided that in order to make an H-R diagram we needed spectral types and *absolute* magnitudes. We also know, however, that the distance to each of the stars in a single cluster is about the same. For the Hyades, all the stars are at a distance of about 46 parsecs (some are 10 percent closer [41 pc] and some are 10 percent more distant [51 pc]. In the case of a more distant cluster like M46, all the stars are within a few parsecs of 1,660 parsecs, so they are all within 0.5 percent of being at exactly the same distance. Let's see what happens if we make an H-R diagram using apparent instead of absolute magnitudes? On such a diagram, we do not know the intrinsic luminosity of any single star, but we do know that if a star in that cluster is, say, five magnitudes brighter in apparent magnitude than another star in the same cluster then it is also five magnitudes brighter in absolute magnitude.

As a result, when we plot spectral types versus apparent magnitudes for all the stars in any single cluster, we will produce a main sequence. A main sequence after all is simply the set of locations on a surface temperature versus luminosity graph for stars of different masses, and a star of a given mass has a unique luminosity and surface temperature combination. We do possess one H-R diagram—for the Hyades—for which the luminosities are calibrated. We also have H-R diagrams for other star clusters, all of which are internally consistent but with the absolute magnitudes for the stars uncalibrated.

Now, imagine that we observe two star clusters, the Hyades and a cluster at an unknown distance. The dwarf (main-sequence) K stars in the second cluster appear 100 times fainter than the dwarf K stars in the Hyades. Since, to within at most a few percentage points (due to age or the elemental composition of the stars' atmospheres), every dwarf K star in the universe is identical to every other dwarf K star, the only reason the dwarf K stars in the second cluster are 100 times fainter than the dwarf K stars in the Hyades is that the second cluster is ten times farther away (using the inverse square law for light) than the Hyades Cluster. We therefore know that the distance to the second cluster is about 460 parsecs.

This method for determining distances to clusters, by comparing the apparent magnitudes of the main-sequence stars in the distant cluster to the absolute magnitudes of main-sequence stars in a star cluster for which we already know the distance (and absolute magnitudes of the stars) is called *main-sequence fitting*. All we need is one star cluster for which the distance is known; from that cluster we can calibrate the absolute magnitudes of the M and K and G and F stars on the main sequence and then use that calibration for all other star clusters we can observe. The Hyades is our starting point. While its distance from the

Sun was first measured by the convergent point method in 1910, it has since been measured in at least a half dozen different ways, all of which agree to within about 10 percent. Even better, the Hipparcos satellite measured the parallaxes to the individual stars in the Hyades, making it possible to determine the absolute magnitude very accurately for each star, which in turn yields extremely accurate absolute magnitudes. Now we have a method to determine distances to any star cluster for which we can plot a main sequence, and we will know the distance to that cluster to within the same degree of accuracy as in our measurements for the distance to the Hyades.

New Astrophysics from H-R Diagrams of Star Clusters

When astronomers began measuring the luminosities and spectral types of stars in star clusters and plotting H-R diagrams for these clusters, several general patterns quickly emerged. All star clusters have main sequences that are truncated at the high luminosity-high temperature end: often the O and/or B and sometimes the A and even F spectral types are missing from the main sequence. On the other hand, the G, K, and M stars are always present on the main sequence (unless the cluster is so far away that the M stars are too faint to be spotted). In addition, the greater the number of hot stars missing from a cluster's main sequence, the more giant stars are present above its main sequence. For many clusters, the main sequence appears to turn sideways

---

Figure 11.6. H-R diagrams (or color-magnitude diagrams) for three star clusters. Top left, NGC 2477. Top right, NGC 188. Bottom, h and chi Persei. The most obvious feature in all such diagrams is the main sequence. H and chi Persei has stars from the top (V=8) to the bottom (V=16) of the main sequence and has no red giants while the other star clusters are missing their hottest, brightest main sequence stars (V < 12 for NGC 2477; V < 15 for NGC 188) and have red giants. Images from Platais et al., *Monthly Notices of the Royal Astronomical Society* (2008) 391, 1482, reproduced with permission from Wiley-Blackwell Publishing (top left); Platais et al., *The Astronomical Journal* (2003) 126, 2922 (top right); Slesnick et al., *The Astrophysical Journal* (2002) 576, 880 (bottom).

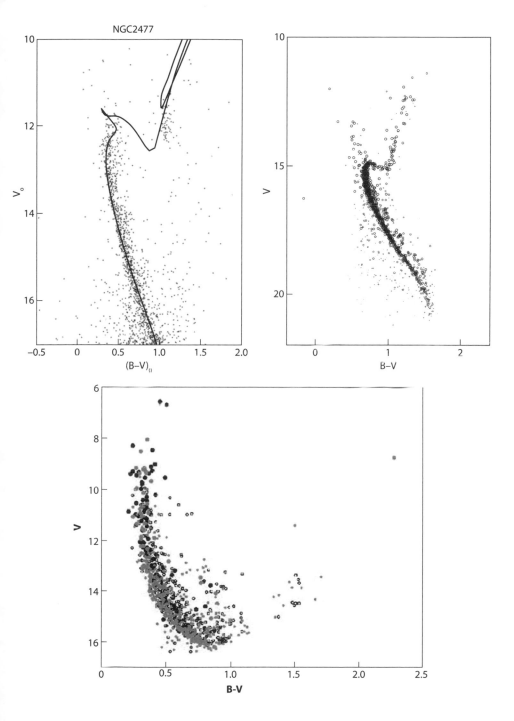

at the high temperature end and then to curve upwards, connecting the main sequence almost seamlessly to the region of the giant stars.

The primary reason that the hottest stars are absent from some cluster H-R diagrams is that stars do not live forever. Remember that the hottest stars are also the most massive; remember also that more massive stars are much more luminous relative to their mass than are less massive stars (the mass-luminosity relationship). The rate at which the most massive stars use up their nuclear fuel is therefore much greater than the rate at which lower mass stars do so, with the result that the higher-mass stars in a cluster have shorter main-sequence lifetimes than the lower-mass stars and will disappear from the main sequence first. Therefore, the main sequences for some star clusters might be missing some of the hottest, most luminous, most massive stars, since these stars have died. A secondary reason for the absence of hot, main-sequence stars in some clusters is that not every cluster gives birth to stars under conditions conducive to forming such stars.

Remember that we concluded that the position of every dwarf star (these being the stars on the main sequence) is determined almost entirely by the mass of that star. We were unable, however, to differentiate between a spectral class K red dwarf and a spectral class K red giant (Hertzsprung could do this with Maury's spectral subclasses, but those subclasses were not included in the spectral types of stars generated by Annie Jump Cannon). The H-R diagrams for star clusters solve this problem: since all the stars in a cluster are, to a very high degree of accuracy, at the same distance from the Sun, we can directly compare the apparent luminosities of all the stars in a cluster, even if we do not know the distances (and therefore the intrinsic brightnesses) of the cluster stars. With a star cluster, we immediately know which stars are dwarf stars: they are the stars that populate the main sequence. Conversely, if one or more K stars in a cluster are 1,000 to 100,000 times brighter than the fainter, main-sequence K stars in that cluster, then we know that the brighter K stars are giants or supergiants.

By comparing the main sequences of star clusters, we can measure distances to one cluster after another, gradually stepping our way farther and farther out into the Milky Way, and even beyond. We can in fact extend our reach out as far as the furthest cluster for which we can resolve the individual stars and measure their apparent magnitudes and spectral types. We are limited only by the reach of our telescopes and our patience.

But there is still one piece of the puzzle missing, and it's an important one. We don't yet know in detail how stars generate their luminosities and temperatures in such a way that every star with the same mass looks the same. Once we have the answer to that question, the H-R diagram (or color-magnitude diagram) will indeed become the magnificent tool we have been seeking, the tool that will allow us to determine the ages of clusters, of the stars in the clusters, and ultimately the age of the universe.

✦ ✦ ✦ CHAPTER 12 ✦ ✦ ✦

Mass Matters

I have reached the conclusion that the absolute bolometric
luminosity of a star depends mainly on its mass . . . .

—Arthur S. Eddington, in "On the Mass-Luminosity Relation,"
Monthly Notices of the Royal Astronomical Society (1925)

In the end, virtually every aspect of stellar astrophysics rests on a single property of stars—mass. In this chapter, we try to make clear exactly why this is so.

The Battle between Gravity and Thermal Pressure

Stars are born regularly in the Milky Way, coming together from fragments of giant clouds in interstellar space in regions like the Orion Nebula. Each interstellar cloud is a volume of space filled mostly with gas (single atoms and molecules) and is characterized by a temperature (or a narrow range of temperatures throughout the cloud), size, mass, composition, and speed of rotation. The temperature of the cloud describes the kinetic energy of the atoms or molecules in the cloud, while the size and mass of the cloud are functions of the cloud's gravitational strength. Since gravity is an attractive force, it naturally acts to compress the cloud. Heat, however, provides a natural pressure that acts to resist gravity and, if possible, to expand the cloud. And rotation provides a mechanism that resists gravity in one direction (perpendicular to the rotation axis) but has no effect on the cloud in the direction parallel to the axis of rotation. These clouds are in short battlegrounds between gravity pulling inwards and thermal pressure pushing outwards, and it is the conflict between these two physical processes that determines the life stories of stars, from birth to death.

Some interstellar clouds are hot and have little mass. In these clouds, expansion wins and stars do not form. If, however, in a sufficiently small volume of space, the mass of the cloud is large enough and the temperatures low enough, gravity wins and stars begin to form. In the latter case, a fragment of the cloud begins to collapse, getting smaller and smaller and smaller. Since the cloud is a gas, and since gases heat

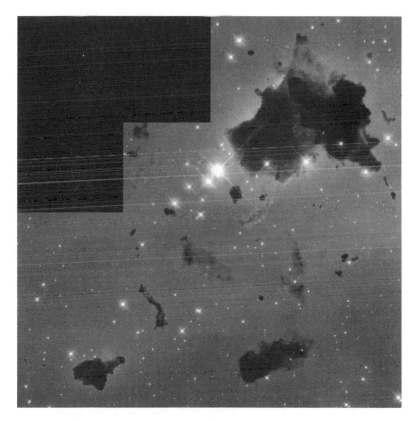

Figure 12.1. Thackeray's globules are dense, opaque clouds of gas and dust in which stars may be forming. These globules are seen silhouetted against nearby, bright, newborn stars in the star-forming region known as IC 2944. Image courtesy of Bo Reipurth (University of Hawaii), NASA, and the Hubble Heritage team (STScI/AURA).

up as their volume shrinks, the contracting cloud fragment begins to heat up. Particles of gas in the warm cloud then radiate away from themselves some of the heat and begin to cool down. Initially, when the cloud is transparent, heat in the form of light can quickly escape from the entire volume of the cloud as soon as it is generated. As a result, in this early stage of star formation, the cloud does not heat up very much and gravity continues to hold the upper hand in its struggle with thermal expansion.

The strength of the force of gravity increases in direct proportion to the magnitudes of the attracting masses but, like the inverse square law for light, decreases in proportion to the square of the distance between the two masses. With the mass of an interstellar cloud fragment squeezed into a smaller volume, the separations between particles become smaller while the masses of these particles remain unchanged. Predictably, as the separations decrease, the self-gravity of the cloud increases dramatically and the cloud squeezes itself even more and becomes even smaller. Initially, the cloud does not warm up much, so gravity quickly gains a huge advantage over thermal expansion and the collapse of the cloud accelerates.

As the cloud is squeezed and becomes denser, it becomes less transparent. Much of the heat that is generated will be radiated away, but over time more and more of the heat will be retained, trapped inside an increasingly opaque cloud. While the now opaque cloud continues to squeeze itself further, it begins to warm up just a little bit more. Once again the battle is joined, with thermal expansion getting a second chance to resist gravity. What we have now is a *protostar*, an object only the outer layer of which can radiate heat into space. Any heat generated deep inside the protostar is trapped inside until it can work its way to the surface. As the center of the protostar gets hotter and hotter, the increasing thermal pressure pushes back ever harder, resisting the compressive force of gravity. For a time, thermal pressure will slow or even halt the gravitational contraction of the protostar, but the protostar continues to radiate heat into space from its surface.

Over long periods of time, as the surface loses heat to space, more heat from the inside of the protostar is transported to the surface; this heat is radiated away and the inside of the protostar cools. Gravity still

Figure 12.2. Star birth clouds in M16. The stars are embedded inside finger-like protrusions extending from the top of the nebula. Each "fingertip" is somewhat larger than our solar system. Image courtesy of NASA, ESA, STScI, and J. Hester and P. Scowen (Arizona State University).

has the advantage. When the inside cools, the protostar shrinks a little, heats up a little, again cools off a little, and again shrinks a little. With each infinitesimal change, both the core and the surface of the star get hotter. Thermal pressure refuses to yield. Yet, the forming star is simply not able to generate enough heat fast enough to hold off gravity. Gravity will continue, inexorably, to squeeze the cloud smaller and smaller unless the collapsing cloud can find a heat source that can generate energy as quickly as the cloud radiates it into space.

It's All about Mass

At some point, if the collapsing protostar is massive enough, the temperature and density in the core will cross a critical threshold at which the nuclear fusion of hydrogen into helium via the proton-proton chain becomes possible. At this moment, the protostar becomes a star, a nuclear fusion machine. It now has its own internal mechanism for generating energy in the form of heat that it can use in resisting gravity. If the core is not hot enough to push back hard enough to stop gravity from squeezing the star into an even smaller volume, then gravity will squeeze it a bit more. The core will become smaller, denser, and hotter, and the rate of nuclear fusion reactions will increase dramatically (because of the increase in temperature). This cycle of compression and increased heating and heat production will continue until the heat released from fusion reactions in the core generates exactly enough pressure to balance the amount of energy lost from the surface to space. When this balance is reached, the battle of compression (gravity) versus expansion (thermal pressure) inside the star reaches an impasse. This period during which the inward pull of gravity and the outward push from heat are in equilibrium, will hold for millions, billions, or even many tens of billions of years, depending on the mass of the star, but all the same the truce is only temporary. Gravity will not give up the fight.

At the moment of equilibrium, the rate of fusion reactions is exactly what is necessary to generate enough heat to push back against gravity and halt the contraction of the star. If the star were to shrink more (if gravity squeezed too much), the temperature in the core would increase. This would cause the rate of nuclear reactions to increase further, in turn releasing even more heat, which would cause the star to expand. As a result of the expansion, the core temperature would decrease, the rate of nuclear reactions and therefore the pace of heat production would decrease, the core temperature would drop, the star would shrink a little bit, and finally the star would ease back into balance. When a star reaches this balance between the compressive force of gravity and the expansive thermal pressure generated by nuclear fusion reactions in its core, the star settles down as a stable object with a constant core temperature, constant surface temperature, and a stable radius. It is now a main-sequence star and will remain so for as long as the conditions for this equilibrium remain in place.

If the mass of the collapsing protostar is below about 8 percent of the mass of the Sun, it will never generate sufficient temperatures and pressures to trigger proton-proton chain reactions, and so it will never become a main-sequence star. If, however, the protostellar mass is below this threshold but above about 1 percent of the mass of the Sun, the internal temperatures and pressures will become high enough to trigger the direct fusion of deuterium into helium, producing an object known as a *brown dwarf*. Since stars have about one deuterium atom for every 6,000 hydrogen atoms, brown dwarfs have very little fuel available (in comparison to stars) that they can tap for nuclear fusion; hence, even the most massive brown dwarfs lack the fuel necessary to power deuterium fusion reactions for more than about 100 million years. Once brown dwarfs run out of deuterium fuel, they slowly fade away.

Newly formed protostars have cooler surfaces than stars and emit very little light. In order to detect and study these faint and cool objects, astronomers make use of large telescopes of two kinds: infrared-optimized telescopes on the ground and in space that can collect the light emitted by warm (infrared) objects and radio telescopes that collect the light (radio waves) emitted by cool objects. When we think about what quantities astronomers measure that allow them to locate stars on the H-R diagram, we discover that protostars do not fit the traditional H-R diagram format. They are too cool and too faint. We can, however, imagine extending our H-R diagram to include both much lower temperatures and much lower luminosities. If we did this, we could place protostars somewhere far beyond the right edge of and far below the faintest red dwarf stars on the traditional region of the H-R diagram.

As protostars become opaque, they become warmer. Thus, they move from right to left (getting hotter) as they near the standard region of the H-R diagram. They also quickly rise to become much brighter than normal stars because, though cool, they are very large and thus have very large surface areas from which to radiate light into space. Thus, we can imagine that forming stars, after first climbing from below and to the right of the main sequence, then approach the main sequence from above and from the right. Now, the H-R diagram has become a tool for tracking the birth of stars.

When nuclear fusion turns on in the core of a newborn star and that star reaches equilibrium, the star settles onto the main sequence. The location on the main sequence where any single protostar will ultimately settle is determined by the mass of that protostar: more massive stars exert greater gravitational compressive forces on themselves so they must become hotter than less massive stars in order to reach gravity-pressure equilibrium. To become hotter, they must generate more energy every second through nuclear reactions in their cores. Due to their size, they are of course also more luminous than lower mass stars.

This *mass-luminosity relationship*, discovered by Eddington in the 1920s, explains why the main sequence is a mass sequence, with the most massive stars also being the hottest and most luminous, while the least massive stars, which need much less heat to counterbalance gravity, are the coolest and least luminous. The mass of the star and the laws of physics choreograph a balance.

The ability of a star to resist gravity via nuclear fusion reactions that convert hydrogen to helium cannot last forever. Eventually the star will run out of unspent hydrogen fuel in its core. What are the main-sequence lifetimes of more massive stars? A 10-solar-mass star has a main-sequence lifetime that is about thirty times shorter than the lifetime of the Sun: it lives a mere 30 million years. A 50-solar-mass star could survive for only about half a million years.

After the hydrogen in its core is exhausted, the proton-proton chain fusion reactions will cease, and the star will no longer be able to replace the heat lost from its surface. Gravity, by patiently outlasting the hydrogen supply, wins again.

···CHAPTER 13···
White Dwarfs and the Age of the Universe

The position of a white dwarf [on an H-R diagram] of given mass and composition depends on its temperature, and hence on its age . . . and a star of a given mass describes, as it cools off, an individual straight line on a Hertzsprung-Russell diagram.

—Leon Mestel, in "On the theory of white dwarf stars," *Monthly Notices of the Royal Astronomical Society* (1952)

The finite main-sequence lifetimes of stars lead us directly toward two different methods for estimating the age of the universe. By the end of this chapter, we will have in hand one of these age estimates; namely, an estimate obtained from the temperatures and cooling times of white-dwarf stars. The second estimate, which derives from observations of large clusters of stars, will be the subject of the following chapter.

Amazingly, it is the inert cores of old, dead stars—for that is what white dwarfs are—that provide the means to estimate their ages from their temperatures. In addition, white dwarfs are fundamental for understanding objects known as "Type Ia supernovae," which are the result of white dwarfs in binary star systems growing too big and then blowing up in catastrophic explosions; and Type Ia supernovae are in turn of paramount importance for understanding the expansion history of the universe and provide yet another method for estimating its age. Clearly then, it would benefit us to look more closely into the astrophysics of white dwarfs.

More than 98 percent of all stars will end their lives as white dwarfs, but that does not mean that 98 percent of all stars have already turned into white dwarfs. The population of white dwarfs depends on stars aging and dying, so astronomers should find lots of white dwarfs to study if the universe is old, but only a few if the universe is young. What, then, is a white dwarf? And how many of them are out there?

Low-mass Stars

The lowest-mass stars, those with masses less than about half that of the Sun, never complete the full set of proton-proton chain reactions. Remember that collisions of atomic nuclei require that particles move very fast, fast enough that positively-charged particles can collide rather than repel. Remember also that temperature is a measure of the average speed of the particles in a gas. In the cores of these stars, the temperatures are high enough for one proton (the nucleus of a hydrogen atom, with a charge of plus one) to collide with another proton in order to form a deuterium nucleus (the nucleus of a hydrogen atom containing one proton and one neutron, also with a charge of plus one) and for a deuterium nucleus to collide with a proton to form a ^{3}He nucleus (containing two protons and one neutron, and thus with a charge of plus two). Hotter temperatures, however, are required in order for two ^{3}He nuclei, each with a net charge of plus two, to collide and fuse into a single ^{4}He nucleus, and these lowest-mass stars lack the gravitational strength to squeeze themselves hard enough to generate such temperatures at their cores. Low-mass stars also do not have cores that are isolated from the process of convection that transports heat upwards through their envelopes to their surfaces. Instead, the upward welling of hot gas in convection cells takes material from the core all the way to the surface, while cooler, denser material plunges from the surface all the way to the center. As a result, in the lowest mass stars nearly 100 percent of the volume of the star ultimately cycles through the core and becomes available for nuclear fusion. Given that the low masses of these stars lead to relatively low surface and core temperatures and that they have low nuclear fusion rates and low luminosities, a star with only one-quarter of the mass of the Sun would have a main-sequence lifetime of over 300 billion years. As we shall see, 300 billion years is about twenty times longer than any estimates of the current

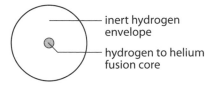

Figure 13.1. The internal structure of a star when it is on the main sequence. In the core, hydrogen fuses to helium. An inert hydrogen envelope surrounds the core. The diameter of the core is about 10 percent of the diameter of the star.

inert hydrogen envelope

hydrogen to helium fusion core

age of the universe. This means that every low-mass star ever born in the universe is still a main-sequence star. Since none of these small stars have died yet and since most stars are small, the universe should have relatively few white dwarfs. This conclusion matches astronomers' observations.

Red Subgiants

In stars with masses greater than about half the mass of the Sun, the steady production of ^{4}He eventually exhausts the supply of hydrogen in their small cores that can be converted to helium. What is more, hot plasma from the core of such a star is not able to convect upwards and mix with plasma from the outer layers of the star. As a result, only hydrogen in the core (which contains about 10 percent of the mass of the star and is about ten percent the diameter of the star) is available for

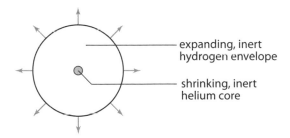

expanding, inert hydrogen envelope

shrinking, inert helium core

Figure 13.2. The internal structure of a star when it leaves the main sequence. In the core, the hydrogen fuel is exhausted and the helium is too cool for fusion. The cooling core contracts, which has the effect of heating the surrounding hydrogen envelope, causing the outer layers of the star to expand. The star is now a subgiant.

fusion. When the core runs out of hydrogen, gravity once again gains the upper hand and the star again begins to contract. The contracting star now consists of two principal parts, a shrinking and inert (no active fusion reactions) helium core and an inert hydrogen envelope. What happens next is one of the most counterintuitive phenomena in stellar astrophysics. The core of the star gets smaller and, as a result, the star as a whole expands.

As gravity crushes the core into a smaller and smaller volume, the temperature of the compressed core begins to rise, from ten to 15 to 20 to 25 million K. While the helium core temperature rises due to gravitational compression, the temperature at the bottom of the hydrogen envelope also must rise because it is heated from below by the rising temperature of the shrinking core. This gradual and quite significant rise in internal temperature forces a parallel rise in internal pressure. As a result, while the very innermost part of the star is squeezed, compressed, and heated by gravity, the outer parts of the star, pushed further outwards by the rising internal heat and pressure, become more rarefied.

As a result of its expanding outer layers, the star now has a larger surface area than before. And by virtue of this larger surface area, it can

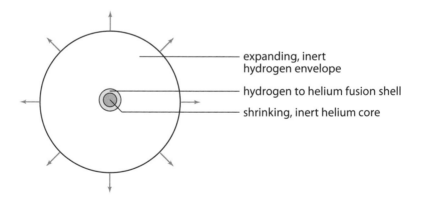

expanding, inert hydrogen envelope

hydrogen to helium fusion shell

shrinking, inert helium core

Figure 13.3. The internal structure of a star when it climbs the red giant branch. The helium core continues to contract. Surrounding that core, a shell of hydrogen becomes hot enough to fuse into helium. The release of heat from the fusion reactions further heats the outer hydrogen envelope, causing it to expand even more. The star is now a red giant.

radiate away the same total amount of energy (or even more) at a cooler temperature. Consequently, if we track changes in the luminosity and surface temperature of an intermediate mass star that has reached the end of its main-sequence lifetime, we find that as stars exhaust the fuel in their cores, they move upwards and to the right on an H-R diagram. They move upwards (become more luminous) because they are hotter in their cores and that heat must be transferred to their surfaces and radiated into space; and to the right (become cooler) because of the expansion of their surface layers. As a result, they creep off the main sequence, growing bigger, brighter, cooler, and redder. They become "red subgiants."

Red Giants

As the core temperature of a star rises into the tens of millions of degrees, the temperature of the hydrogen gas at the bottom of the envelope just outside the core rises just above 10 million degrees. Gradually, the proton-proton chain turns on in a thin shell of hydrogen gas deep inside the star but outside of the inert helium core. This extra release of energy in a layer outside of the core heats up and expands the envelope further, making the star even bigger, more luminous, and cooler. The star climbs upwards in the H-R diagram and becomes a "red giant." Once again, the star achieves an internal pressure equilibrium, but the balance between gravity and thermal pressure that a star reaches during its red giant stage will be short-lived—only about one tenth as long as the equilibrium reached in its main-sequence lifetime —as less fuel is available for fusion in this thin shell than was available in the core.

Electron Degeneracy

Deep inside intermediate-mass stars that have evolved into red giants, the material undergoes a strange quantum mechanical transition from normal matter to *electron degenerate matter*. This phenomenon of electron degeneracy is important not only for explaining the behavior of intermediate-mass stars that become red giants but also for understanding white dwarfs. Electron degeneracy has to do with two facts: each and every electron is indistinguishable from every other electron; and quantum mechanics limits the energies of all subatomic particles. For example, the electrons in bound orbits within atoms cannot organize themselves into orbits characterized by any arbitrary amount of energy. Instead, their orbits are restricted to specific, quantized levels; this is why each atom has a unique spectral signature in the form of emission lines and absorption lines. The rules of quantum mechanics dictate that two identical particles located in the same small volume of space cannot have the same exact properties (energy, quantum mechanical spin, direction of motion). These rules also limit the total possible sets of properties, or states, available to all the particles. As a result, if in one place inside a star all the low energy states are filled by electrons, another electron cannot come along and occupy a low energy state in that same volume of space. The additional electron either is excluded from this volume of space or it has to remain at a higher energy level whose states are not filled. This uncollegial pattern of behavior, which somewhat resembles a game of musical chairs, is known as the *Pauli exclusion principle*.

Unlike a regular game of musical chairs, however, the players who do not get chairs are not sent home; they continue to mill around the filled chairs (roaming players are in higher energy states than seated players), filling the rest of the room and preventing other potential players from even entering the room. When the gas at the center of a star is subjected to enormous pressure, it becomes degenerate. The gravitational pressure from the outer layers of the star prevents degenerate electrons from leaving the degeneracy region, just as a line of would-be players might crowd the doors trying to get into the musical-chairs room and prevent those players without seats from leaving the room. If our game of musical chairs became a marathon, dragging on for days, our seatless players would become exhausted; they would walk more slowly; they would desperately want to sit down. But with no chairs available, they would be forced to stand, to wander around the room. The human pressure filling this room would no longer depend on the energy of the players but on the mere presence of too many players in too small a place. A degenerate gas builds up an analogous kind of pressure.

The phenomenon of electron degeneracy develops at the extreme pressures and densities found in the centers of stars, but degenerate electrons exist in normal metals on Earth as well. They are responsible for the high electric and thermal conductivity of metals, and their presence makes metals, like white dwarfs, very difficult to compress. In fact, a good analogy to a white dwarf would be a ball of liquid metal. Qualitatively, the difference between a white dwarf and the liquid metal ball is the enormous density and pressure of the white dwarf, so that there are more degenerate electrons in a given volume in the white dwarf than in the same volume of the liquid metal ball.

Just below the surface of a star that becomes a red giant, the temperature rapidly climbs above 30,000 K. Below this level, not only are all the hydrogen atoms ionized but all the helium atoms in the star are so hot that both electrons are stripped from their nuclei. Thus, virtually the entire volume of the star is filled with fully ionized hydrogen and helium nuclei, embedded in a sea of electrons that are no longer bound to any nucleus; instead, the free electrons are able to travel throughout the volume of the star. At the high densities near the centers of these stars, the free electrons are squeezed into such a small volume that they eventually fill up all the available low energy states— all the chairs are taken. Gravitational pressure squeezes even more electrons into this small volume. The electrons, given their temperatures, would prefer to occupy low energy states, but the Pauli exclusion principle prevents this from happening. The electrons cannot leave this volume, so they must remain at higher energies than they would normally have if the lower energy states were open. Consequently, the extra electrons, hot and confined, push back against gravity.

As a result, the pressure that resists gravity is no longer thermal pressure; that is, it is no longer determined by temperature. Instead, the pressure comes from electrons that cannot settle down to energy levels appropriate for the temperature, since all of those energy states are filled. This pressure is known as *degeneracy pressure*. Because free electrons conduct energy very quickly, and because the entire volume of the degenerate core of an intermediate-mass star is filled with free electrons, any temperature differences in the core are almost instantly smoothed out and these stellar cores quickly become nearly isothermal (having the same temperature throughout).

A Carbon/Oxygen Core

When the temperature of the core reaches an incredible 100 million K, helium nuclei are moving fast enough to collide with each other. When the density of the core surpasses 10^4 grams (10 kilograms) per cubic centimeter, the helium nuclei can no longer avoid collisions. The nuclear fusion of helium nuclei into carbon nuclei begins.

The fusion process occurs in two steps. First two ^{4}He nuclei collide; they stick together and form a beryllium-eight (^{8}Be) nucleus. Some mass is converted to energy and this energy is released as a gamma ray. The ^{8}Be nucleus, which contains four protons and four neutrons, is very unstable, (^{9}Be, which is formed during the rapid collisions of particles during supernova explosions, has four protons and five neutrons and is stable, but normal stars make ^{8}Be, not ^{9}Be), and it quickly falls apart again (the half-life of ^{8}Be is about a billionth of a billionth of a second) into two ^{4}He nuclei. At temperatures of 100 million K and above, however, the rate at which ^{8}Be can form is comparable to the rate at which it falls apart, so some ^{8}Be nuclei survive long enough to participate in the next step in this fusion process, in which a third ^{4}He nucleus collides with a ^{8}Be particle to form a carbon-twelve (^{12}C) nu-

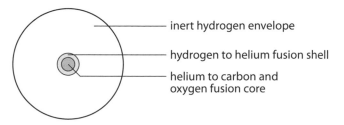

inert hydrogen envelope

hydrogen to helium fusion shell

helium to carbon and oxygen fusion core

Figure 13.4. The internal structure of a star in which the core has become hot enough for the helium to fuse into carbon and oxygen. The star is now a horizontal branch star.

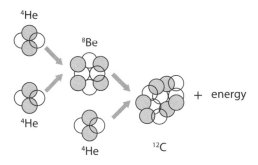

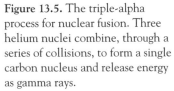

Figure 13.5. The triple-alpha process for nuclear fusion. Three helium nuclei combine, through a series of collisions, to form a single carbon nucleus and release energy as gamma rays.

cleus, containing six protons and six neutrons. Because, for historical reasons, the helium nucleus is referred to by physicists as an alpha particle, and because this process ultimately turns three alpha particles into a single carbon nucleus, it is known as the *triple-alpha process*. Along with the triple-alpha fusion process, intermediate-mass stars turn a little bit more mass into energy by producing a small amount of oxygen through a reaction in which a ^{4}He nucleus collides and combines with a ^{12}C nucleus to create an oxygen-sixteen (^{16}O) nucleus. Slowly but surely, the cores of these stars fill with helium, carbon, and oxygen, with some of the carbon and oxygen mixing upwards into the outer layers of the star.

When these stars die as planetary nebulae, an event we will discuss later in this chapter, much of the carbon and oxygen they generate in these reactions will be expelled into space. The exposed carbon/oxygen core will be left behind and will become a white dwarf, while the expelled atoms will become parts of molecules and dust grains that will collect into interstellar clouds and then form into the next generation of stars and planets. In this way, each generation of red giants enriches the galaxy with greater amounts of elements that are heavier than hydrogen and helium. Ultimately, all of the carbon atoms on earth, whether in coal, graphite, diamonds, organic molecules, or carbon dioxide, were manufactured in the cores of red giant stars. And so was all the oxygen that we breathe.

When the triple-alpha process commences, energy is released in the core. Normally, a release of energy would increase the temperature of the gas and the higher temperature would increase the pressure, which would cause the core to expand. However, because the core is electron degenerate, the pressure is controlled by the degenerate electrons, not by the temperature. As a result, the release of energy from a few triple-alpha process reactions raises the temperature, but the core does not expand. As the core temperature rises, more fusion reactions occur more quickly and the core becomes hotter still.

Finally, as a result of the climbing temperature, new energy states become available to the electrons and the electrons are able to occupy these newly available non-degenerate states. In our analogy, many more chairs have been added to the musical chairs game. All players now have seats, and there are some empty ones left over. The condition of degeneracy has been removed. Suddenly, the core can behave like a normal gas, with thermal pressure balancing gravity. On the H-R diagram, the onset of the triple-alpha process occurs at the highest point on the red-giant branch. After core fusion of helium into carbon begins, these stars contract a little bit and settle down to much lower luminosities though with slightly higher temperatures. On the H-R diagram, they land on the horizontal branch, which is located above the main sequence and to the left of (hotter than) and below (less luminous than) the region of red giants. The star begins a slow, steady

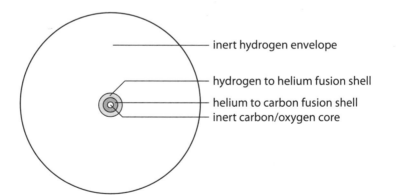

Figure 13.6. The internal structure of a red supergiant star. At the center, an inert core of carbon and oxygen is being deposited as the residue of the fusion of He. This core will be left behind as a white dwarf, after the star expels its outer layers in the planetary nebula phase.

phase of its life as a helium-to-carbon fusing red giant. Gravity must once more be content with a draw, albeit only temporarily.

The internal structure of these aging, intermediate-mass stars begins to resemble an onion. They have inert central cores in which newly formed carbon nuclei are accumulating. Surrounding the mostly inert carbon cores are shells of helium in which helium-to-carbon fusion occurs; surrounding these are thin shells in which hydrogen fusion is generating more helium. The outermost layers are inert shells of hydrogen in which the hydrogen is too cool for fusion. As they age, intermediate-mass stars once again expand and become more luminous, ultimately becoming red supergiants once again.

The Instability Strip

During the horizontal branch phase in the life of a star, the star goes through periods during which the outer layers of the star heat up and consequently expand and cool off. At the lower temperatures in these outermost layers, some electrons are able to combine with atomic nuclei or ions to form less highly charged ions or even neutral atoms. The enlarged star is more efficient at emitting light into space, in part because it has more surface area from which to emit light and in part because the neutral atoms and less highly charged ions are more transparent to outgoing light from lower layers than were the free electrons and more highly charged particles. Consequently, the outer layers of the giant star cool off so much that they eventually lack the thermal pressure needed to continue in their inflated state, whereupon the outermost layers of the star fall inwards again and the cooling envelope shrinks back to the size at which thermal pressure and gravity should reach equilibrium. When the star reaches that size, however, the envelope is still falling inwards. It cannot throw on the brakes quickly enough; it overcompensates and contracts too far. As a result, rather than coming back into balance, the envelope gets too small and heats up too much. The additional heat generates enough thermal pressure to finally slow down and stop the contraction. But now, the neutral atoms have turned back into ions, which are less transparent to outgoing radiation than were the neutrals. Once again the outer layers of the star heat up and puff out. The star expands back to the size at which gravity and thermal pressure should balance, but this time it cannot throw on the brakes quickly enough to stop the outward moving atmosphere from expanding past the equilibrium point. The star is now unstable and has no choice: another cycle of expansion-cooling-contraction-heating begins.

These cycles of pulsation by giant stars, during which they alternately become bigger and brighter and cooler and then smaller and fainter and hotter are manifestations of the imbalance between inward gravitational pull and outward thermal push. Unable to achieve pressure equilibrium, the stars continue to pulsate. Certain pulsating stars called Cepheid variable stars, which we will visit in Chapters Fifteen and Sixteen, provide yet another means for estimating the ages of stars, the galaxy, and the universe.

Planetary Nebulae

One of the consequences of pulsation is that a star can expand its outer envelope so rapidly that gravity cannot slow down the outward rushing shell of gas enough to stop it and drag it back down into yet another pulsation cycle. The envelope escapes from the star, like a smoke ring puffed off into space. All that is left behind is the extremely hot and small core of the red giant, which is made almost entirely of helium, carbon, and oxygen. Arthur Eddington named these remnants *white dwarfs*.

Red giant stars are very effective at puffing off these rings at speeds of thousands of kilometers per second and at mass-loss rates of up to one hundred-thousandth of a solar mass per year. At that rate, in only 100,000 years a star could eject into space as much mass as is contained in the entire Sun. In less than one million years, a star that is born with nine times the mass of the Sun could shed 90 percent of its mass. Through this mechanism red giant stars, whose main-sequence progenitors were greater than roughly half a solar mass and less than about nine solar masses, can and will shed most of the mass with which they

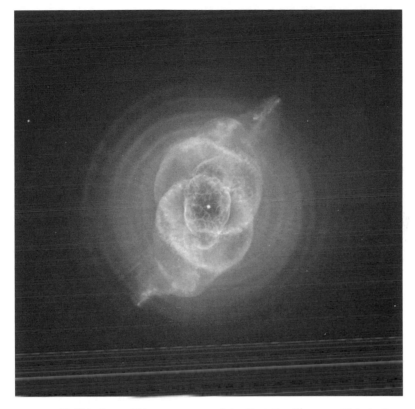

Figure 13.7. Hubble Space Telescope image of the Cat's Eye Planetary Nebula. Image courtesy of NASA, ESA, HEIC, and the Hubble Heritage team (STScI/AURA).

were born; a star that begins its main-sequence lifetime with a mass five times greater than that of the Sun likely will loft more than four solar masses into space and end up with a mass only about one-half to three-fifths that of the Sun. In fact, virtually every star in the mass range from one half to about nine solar masses will end up shedding all but about half a solar mass during this phase.

While these dying stars are actively expelling their outer layers into space, they are called *planetary nebulae*. (The name comes from the fact that in the telescopes available to astronomers 200 years ago, these objects appeared big and circular, looking very much like fuzzy planets and very unlike stars.) At temperatures of up to 150,000 K, the left-behind cores are white hot and emit far more ultraviolet than visible

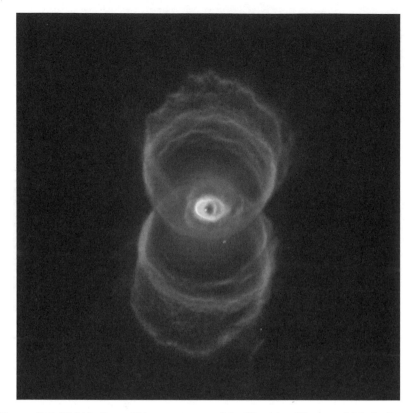

Figure 13.8. Hubble Space Telescope image of the Hourglass Planetary Nebula. Image courtesy of Raghvendra Sahai and John Trauger (JPL), the WFPC2 science team, and NASA.

light. The ultraviolet light heats up the escaping planetary nebula gases and causes the nebulae to glow.

White Dwarfs

As we have seen, white dwarfs are stellar remnants, the abandoned cores of red giants. They are bizarre objects, extremely remote from anything we experience in everyday life. On the H-R diagram, they appear at the far left, because of their high temperatures, and far below the main sequence, because of their small surface areas (with about

10,000 times less surface area than the Sun, they are about the same physical size as the Earth) and consequent small luminosities.

In Chapter Nine, we saw how the observed temperature and luminosity of Sirius B, in combination with the Stefan-Boltzmann law, could be used to demonstrate that Sirius B was extremely small, with a diameter nearly identical to that of the Earth. We also know from observations of the Sirius/Sirius B binary system that the mass of Sirius B is almost identical (98%) to that of the Sun. We thus have an object the size of the Earth but with 300,000 times more mass. This combination of mass and size means that the average density of Sirius B must be 300,000 times greater than the average density of the Earth, upwards of one ton per cubic centimeter, and that the gravitational force at the surface of Sirius B must be 300,000 times greater than the gravitational force at the surface of the Earth. At such high densities and pressures, almost the entire volume of a white dwarf (99 percent of its mass) is filled with degenerate electrons, just as was the core of our intermediate-mass star shortly before the onset of the triple alpha process. A white dwarf, then, is an earth-sized sphere comprised of helium, carbon, and oxygen nuclei that is filled with a fog of degenerate electrons. The pressure from the electrons helps the white dwarf maintain a nearly constant density throughout almost its entire volume. The temperature is nearly constant as well, because degenerate electrons being excellent heat conductors, if one part of the inside of a white dwarf were to be much hotter than another, the degenerate electrons would very quickly smooth out any significant heat differences. The only deviation from this near-isothermal, degenerate electron nature of the white dwarf is to be found in its outermost layers. The surface consists of a very thin, opaque, insulating layer of non-degenerate helium surrounded by, for about 75 percent of all white dwarfs, a similar layer of hydrogen.

The radius of a white dwarf is determined by electron degeneracy pressure, not by its internal temperature. In turn, the degeneracy pressure depends on the mass but is independent of the temperature of the white dwarf. No matter how much energy the white dwarf radiates into space, no matter how cool it gets, *a white dwarf will stay the same size*. Gravity has forced the state of degeneracy on the white dwarf but cannot further compress the star. Finally, and permanently, gravity has lost the battle.

An isolated white dwarf will maintain its mass and size forever. But a white dwarf will not maintain a constant surface temperature. And therein lies the key we have been seeking. For although not much is left behind when a star blows 90 percent or more of the mass in its outer layers into space, what little is left behind is going to tell us the age of the universe.

White-Dwarf Cooling Curves

When they are first born, white dwarfs are the exposed 150,000 K cores of red giants. Despite their enormous temperatures and pressures, they are not hot and dense enough in their cores to generate heat through the nuclear fusion of carbon and oxygen into heavier elements. In addition, degeneracy pressure keeps them at a constant size so they no longer can generate energy from gravitational collapse or shrinkage. Since they are much hotter than interstellar space, they must continue to radiate heat, but because they are no longer nuclear fusion machines, they have no sources of heat from which they can replenish the energy lost to space.

So what happens to white dwarfs? They must cool off, and they must do so without shrinking. The rate at which a white-dwarf-sized object cools depends on its luminosity, which in turn depends on its size (total surface area) and surface temperature. Since an individual white dwarf does not change in size as it cools, the hotter it is the more effectively it radiates heat to space. At first the dwarf cools rapidly, but the cooler it becomes the slower its rate of cooling. When (after only about 30 million years) a white dwarf reaches temperatures below about 25,000 K, the cooling rate slows down dramatically.

Let's restate this cooling scenario for white dwarfs. An astronomer finds a white dwarf and measures its temperature and luminosity. The luminosity is determined by the temperature and radius, so from the luminosity and temperature measurements we have effectively measured the radius. In turn, the radius is determined by the mass; we therefore also have measured the mass of the white dwarf. If we know the temperature, radius, luminosity, and mass of the white dwarf today, we know the rate at which it is radiating energy to space, and so we also know what the luminosity and temperature of the white dwarf will be tomorrow, next week, next year, or in a thousand, million, or billion years. Similarly, if we know the luminosity and temperature of the white dwarf today, we know what its luminosity and temperature were yesterday, last week, last year, and even a thousand, million, billion, or 10 billion years ago.

On an H-R diagram a white dwarf follows a cooling curve downward (as it becomes less luminous) and to the right (as it becomes cooler). For an object that cools off but maintains a constant size, such a cooling curve is a straight line on an H-R diagram (but a curved line on a color-magnitude diagram), with white dwarfs of different masses lying on parallel straight lines.

Astronomers are rarely able to see white dwarfs when they are very young and hot, because this phase is so brief. Only a few such objects, with luminosities one hundred to one thousand times greater than the Sun's have been spotted. Most white dwarfs are first observed in the middle-left part of the H-R diagram, with temperatures of 25,000 K to 35,000 K and with luminosities about one-tenth that of the Sun. As they age, white dwarfs slide down the cooling curves toward the lower-middle part of the diagram, and as they slide down, they slide ever more slowly. The entire set of cooling curves for white dwarfs falls well below the main sequence.

Most details of white dwarf cooling are fairly well understood, but making the observational measurements that are needed to determine which cooling model is appropriate for a given white dwarf can be tricky. For example, white dwarfs with thin atmospheric layers of hydrogen will cool at slightly different rates than white dwarfs with thick hydrogen atmospheres, but observers cannot easily determine whether a given white dwarf has a thick or a thin atmosphere. In addition, the composition of the core of a white dwarf determines how much heat it holds and how fast it cools. A white dwarf core dominated by oxygen holds less heat and cools faster than a white dwarf core dominated by carbon. But astronomers have not figured out yet how to accurately

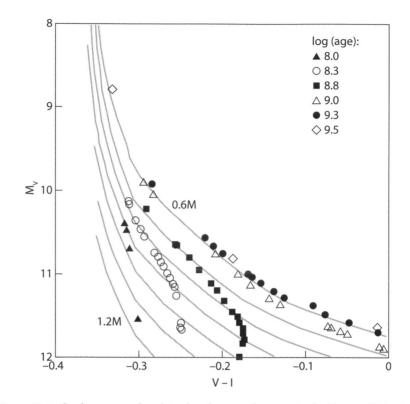

Figure 13.9. Cooling curves for white dwarfs on a color-magnitude diagram. This plot shows only the white dwarf section of the diagram. The lines show the cooling curves for white dwarfs with masses ranging from 0.6 solar masses (top curve) to 1.2 solar masses (bottom curve). White dwarfs with different radii (i.e., masses) can have the same temperature (i.e., V-I color) and thus will have different luminosities. Note that, counterintuitively, more massive white dwarfs are smaller and therefore less luminous than less massive white dwarfs. *Any single white dwarf has a fixed mass and will slide down a cooling curve as it ages, becoming fainter (larger in absolute magnitude) and cooler (with less negative V-I color).* Any single star cluster of a given age (see symbols) would have white dwarfs representing a range of masses. From Jeffrey et al., *The Astrophysical Journal* (2007): 658, 391.

determine the exact composition of a white dwarf core. Given any single white dwarf for which we knew the composition of the atmosphere and core, we could calculate to a high level of accuracy the cooling rate of that white dwarf, for we know that it will cool off very slowly and very dependably, like clockwork. But for a real white dwarf, one whose temperature, luminosity, diameter, and composition are known only to a limited level of accuracy, it is not always obvious which detailed model applies. The best we can do is to calculate either a range of possible ages or an age that is accurate to within a billion years.

The Age of the Universe from a Shortfall of Faint White Dwarfs

Because astronomers know from observations the temperature and luminosity range in which all white dwarfs are born, and because the cooling rates for white dwarfs are well understood, it is possible to determine the age of a white dwarf—it is actually the length of time since it became a white dwarf that is being reckoned—by measuring its luminosity and surface temperature. So what might we expect in our observations of white dwarfs?

The faintest and coolest white dwarfs along any single cooling track must be the oldest white dwarfs. All we need to do is survey the heavens for white dwarfs, find the faintest and coolest ones, and derive the lengths of time they have been white dwarfs from our knowledge of the rates at which white dwarfs cool. If the universe is extremely old, some white dwarfs will be extremely faint and cool. If the universe is extremely young, no white dwarfs will have had enough time to cool off and fade away to obscurity.

Making the necessary observations is easier said than done, however, because white dwarfs are intrinsically very small and cool and therefore very faint and hard to find. The faintest and coolest known have luminosities about 30,000 times fainter than the Sun's and temperatures of about 4,000 K. Such objects are difficult to detect even if they are near the Sun. In order to find them, astronomers need large telescopes to collect enough light. Since white dwarfs are so cool, astronomers also need sensitive detectors that work at very red or infrared wavelengths. And finally, because of the rarity of these oldest and coolest white dwarfs, they need the ability to observe and analyze data from large areas of the sky in order to improve their chances of finding one. For decades, astronomers had no hope of detecting white dwarfs whose

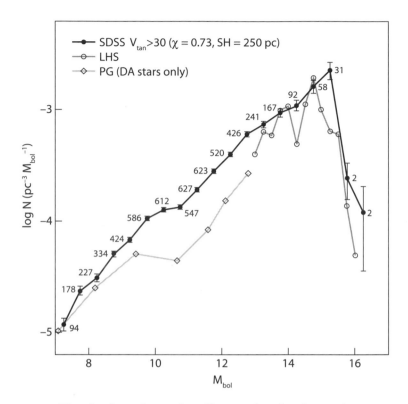

Figure 13.10. This plot shows the number of known white dwarfs per volume segment of the galaxy at different brightnesses (y-axis) plotted as a function of their brightness (x-axis) observed as part of the Sloan Digital Sky Survey. The dashed and dotted lines indicate white-dwarf measurements from other observing projects. Brightness is measured in units of *bolometric magnitude* (or M_{bol}), which indicates the total amount of energy emitted by the white dwarf at all wavelengths, from X-rays to radio waves. The faintest stars are to the right, the brightest to the left. The number of white dwarfs is measured as stars per cubic parsec at each bolometric magnitude. On the plot, the numbers indicate the total number of white dwarfs observed that contribute to each data point. For example, these data indicate that 586 white dwarfs are known with M_{bol} = 10. The volume of the galaxy surveyed by the SDSS to find these white dwarfs indicates the presence of about one white dwarf of this brightness in every 10,000 cubic parsecs (equivalent to a cube just a bit larger than 20 parsecs along each edge). The brightest white dwarfs (small M_{bol}) are rare, less than one in every 100,000 cubic parsecs (a cube with edges of about 50 parsecs in length). Up to a limit (M_{bol} = 15.5), the number of white dwarfs per unit volume of the galaxy increases as the white dwarfs get fainter. The observations, however, show the dearth of very faint white dwarfs: only four white dwarfs are known with bolometric magnitudes larger than 15.5, and the number of such white dwarfs per unit volume of the galaxy drops precipitously. From Harris et al., *The Astrophysical Journal* (2006): 131, 571.

luminosities were less than 30,000 times fainter than the Sun's. They simply lacked the necessary combination of large telescopes, sufficiently sensitive detectors that could also measure light from a large part of the sky at once, and powerful computing hardware. No super-faint white dwarfs were known, but one could argue that they were escaping detection because of our technological limitations.

With the introduction of the current generation of telescopes with more sensitive detectors, and with the advent of modern computing technologies, astronomers now have the ability to detect white dwarfs that are much less than 30,000 times fainter than the Sun and cooler than 4,000 K. The Sloan Digital Sky Survey provides the most sensitive data yet for white dwarfs, and these data show that faint white dwarfs simply are not out there. They do not exist. The stars in the Milky Way are too young. The universe must be younger than about 15 billion years.

The best estimates for the lengths of time that the faintest and coolest white dwarfs have spent on the white-dwarf cooling track are in the range from 8.5 to 9.5 billion years. Given observational uncertainties, a fair assessment of their ages would be between eight and 10 billion years. It would seem that we are well on our way to putting an upper limit on the age of the universe—if, that is, our local neighborhood in the Milky Way, is representative of the whole.

The Age of the Universe from the Population of White Dwarfs in the Solar Neighborhood

Since white dwarfs are intrinsically very faint, virtually all observed white dwarfs are near the Sun. Their ages therefore provide a firm upper limit for the age for the population of stars that make up the disk of the Milky Way Galaxy. Since these stars had to form and live out their main-sequence lifetimes before becoming white dwarfs, the age of the Milky Way is the sum of the main-sequence lifetimes of the stars that became white dwarfs added to the white-dwarf lifetimes of at least 8 to 10 billion years. The oldest white dwarfs are the ones that would have become white dwarfs first, and since these white dwarfs would have come from the most massive stars that have already gone through the red giant to planetary nebula to white-dwarf stages, we can readily estimate a reasonable main-sequence lifetime of about 300 hundred million to 1 billion years for the progenitor stars of this white-dwarf population.

Our next task will be estimating the length of the interval from the beginning of the universe until the first stars formed in the Milky Way. A popular guess, though this is only a guess, is 1 or 2 billion years. Thus, the white-dwarf population near the Sun suggests that the universe is about 9.5 to 13 billion years old.

The Age of the Universe from the Coolest Individual White Dwarf

The Milky Way has two main components, the disk and the halo. Disk stars—our Sun is one—move more slowly than halo stars and generally travel in similar orbits around the center of the Milky Way. Halo stars have larger space velocities and do not share the general "around the center" motion of the other stars in the solar neighborhood. A mere 28 parsecs from the Sun there is a much-studied white dwarf named WD 0346+246. WD 0346+246 is cool and has a high, not-around-the-center space velocity and so is thought to be from the halo of the Milky Way, just passing through our neighborhood.

Astronomers who study the formation of galaxies like the Milky Way believe halo stars formed before disk stars. If this is this case, then WD 0346+246 could be older than the other white dwarfs in our neighborhood. The data suggest that it might be, but not by much. WD 0346+246 has a temperature of only 3,780 K and a luminosity that is 30,000 to 70,000 times fainter than the Sun. Given these figures, cooling curve calculations suggest that the age of this white dwarf is about 11 billion years, which could be older but not significantly older, than the other white dwarfs in our neck of the Milky Way. Again, adding a few hundred million years for the main-sequence lifetime of the progenitor star and 1 or 2 billion years after the birth of the universe but before the birth of any stars, we would conclude from our measurements of WD 0346+246 that the universe must be about 12.5 to 14 billion years old.

The Age of the Universe from White Dwarfs in Globular Clusters

Globular clusters are systems of hundreds of thousands of stars that formed very nearly contemporaneously, in a single burst of star formation early in the history of the galaxy and perhaps the universe. Because so many stars formed so near each other, the gravitational attraction of each star for every other star in a cluster holds a cluster together over the vast ages of universal history. In such a cluster, the first stars to die would be the most massive, which, exploding as they die, would leave behind neutron stars and black holes, but not white dwarfs. Eventually, as intermediate-mass stars begin to die after a few hundred million years, remnant white dwarfs become an increasingly populous subpopulation within a cluster (in Chapter Fourteen, we will discuss how we establish the actual ages of globular clusters). Assuming that globular clusters are among the very oldest structures in the galaxy, white dwarfs in globular clusters are very likely among the oldest white dwarfs in the galaxy.

In 2002, a team using the Hubble Space Telescope identified perhaps the oldest population of white dwarfs yet found in the Milky Way in the closest globular cluster to the Sun (M4). These white dwarfs have ages of 10 to 12 billion years. Again, some time must have elapsed from the formation of the universe until the formation of the first stars in the Milky Way. In addition, a few hundred million years may have passed after the birth of stars in M4 and before the first white dwarfs formed in this cluster. Therefore, the age-dating of this cluster yields a range of ages for the universe of about 11.5 to 13.5 billion years. In 2007, the same team reported on similar Hubble Space Telescope observations, this time of the second closest globular cluster, NGC 6397. For this cluster, they calculated that the white-dwarf cooling sequence age is about 11.5 billion years, which is essentially the same as the age of the white dwarfs in the M4 cluster.

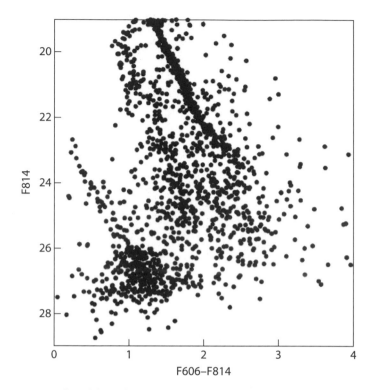

Figure 13.11. A plot of the colors of stars (x-axis) vs. the brightness of stars (y-axis) in the globular cluster M4, in data obtained from the Hubble Space Telescope. This plot is a color-magnitude diagram. The upper dark line, extending from the top-middle of the graph downwards and to the right, is the main sequence; the lower dark line, extending downwards from almost the left edge and at F814 = 23.5, is the white-dwarf cooling curve. In between, most of the scattered points mark out a main sequence of background stars unrelated to M4. The lower (faintest) limit to the white dwarfs seen at about magnitude 27.5 indicates the location on the plot of the coolest and oldest white dwarfs in this cluster. Their ages are calculated to be between 10 and 12 billion years. The x-axis is plotted in units labeled "F606–F814." These are color filters used in Hubble Space Telescope observations that correspond, approximately, to green (F606) and red (F814). From Hansen et al., *The Astrophysical Journal Supplement* (2004): 155, 551.

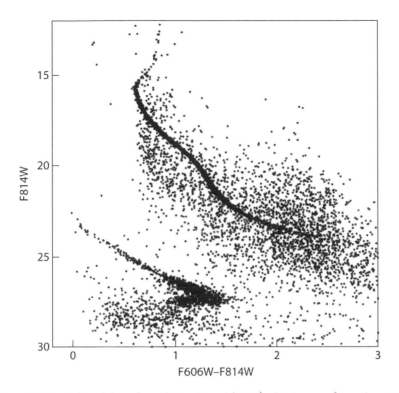

Figure 13.12. A plot of the colors of stars (x-axis) vs. the brightness of stars (y-axis) in data obtained from the Hubble Space Telescope for the globular cluster NGC 6397. The ages of the oldest, faintest white dwarfs in NGC 6397 are calculated to be about 11.5 billion years. From Hansen et al., *The Astrophysical Journal* (2007): 671, 380.

The Age of the Universe from Observations of White Dwarfs

The temperatures, luminosities, and cooling rates for white dwarfs give us, directly, the ages of these white dwarfs. To those ages, we can confidently add a few hundred million years to account for the main-sequence lifetimes of the stars that lived and died and left the dwarfs behind. Sum the figures and we can confidently say that stars were born in the Milky Way more than 11 billion years ago and possibly as much as 13 billion years ago. We can also say, with confidence, that if the universe were 15 billion years old or older, we should observe many white dwarfs that are cooler and fainter than any we have observed. Our telescopes and detectors are sensitive enough to detect such super-faint and cool white dwarfs, but they are nowhere to be found. The universe simply is too young. White dwarfs thus put both a lower and an upper limit on the age of the universe.

They tell us that the age of the universe is significantly older than the 4.5 billion year age of the Sun, the Earth, the Moon, and the oldest known meteorites. In fact, with ages of 11 to 14 billion years, the old-est white dwarfs are nearly three times older than the Sun and Earth, making our solar system a relative youngster.

✦ ✦ ✦ CHAPTER 14 ✦ ✦ ✦

Ages of Globular Clusters and the Age of the Universe

The remarkable similarity between the relative ages of globular clusters, together with their absolute value near 10×10^9 years, has well-known consequences for the history of the Galaxy and the Universe.

—Allan Sandage, in "Main-Sequence Photometry: Color-Magnitude Diagrams and Ages for the Globular Clusters M3, M13, M15, and M92," *The Astrophysical Journal* (1970)

We know that stars are born out of giant interstellar clouds, that some of those clouds form clusters with only a few dozen or few hundred stars while others form clusters with hundreds of thousands of stars. We also know from observations of clusters of newborn stars that the time required for all of the stars in a cluster to form—from the birth of the first star to the birth of the last—is at most a few million years. Except in the very youngest clusters, the age of the cluster (hundreds of millions to many billions of years) is much less than the total length of time (a few million years) for all the stars in a single cluster to be born. Therefore, except for the youngest of newborn clusters, we can very reasonably treat all the stars *in any single cluster* as having the same age.

Open Clusters and Globular Clusters

In most open clusters, the total gravitational pull generated by each star in the cluster on every other star in the cluster is small, too small for the stars to remain gravitationally bound to each other for long. Thus, after only a few hundred million years (astronomers have curious notions about what constitutes a short amount of time) the random motions of the stars in an open cluster will cause such a cluster to expand. The fastest moving of the stars will escape from the cluster, reducing its total mass so that it has an even harder time holding itself together. Inevitably, other stars will continue to slip away from the cluster until the cluster disperses completely.

On the other hand, the large clusters known as globular clusters contain enough stars and enough total mass for gravity to bind those stars to one another forever. It is conceivable that a single star might, through random interactions with other stars, gain a velocity so large that it could escape from the cluster into interstellar space. But the velocity needed for such an escape is sufficiently large that very few stars in globular clusters ever achieve it. Even after many billions of years, almost all globular-cluster stars remain bound to the clusters in which they were born.

The basic understanding that gravity keeps some clusters together forever, while insufficient gravity permits other clusters to disperse in much less time than the life spans of the stars themselves, leads us to ask a simple but pivotal question: Which type of cluster should we observe if we want to find very old stars?

Clearly open clusters are out. They are young and, with very few exceptions, all the stars in them must be young as well. While we cannot yet (but by the end of this chapter will confidently) guarantee that the stars in globular clusters are old, we certainly know that they could be. So we will turn our attention now to globular clusters.

Isochrones: Physical Age versus Life Cycle Age

While we must keep in mind that all the stars in any single cluster are of very nearly the same physical age, we also need to remember that the rates at which stars use up their nuclear fuel and move off the main sequence to become red giants and eventually white dwarfs depend on the masses with which the stars were born. More massive stars exhaust their fuel supplies for nuclear fusion reactions, move off the main sequence, and die long before their lower-mass siblings. Thus, the *age of a star* is not equivalent to *how far along that star is in its life cycle*. A single cluster might contain stars at every conceivable life cycle stage—many dwarfs still on the main sequence, a few subgiants, a handful of red supergiants, a small number of white dwarfs—even though all the stars in that cluster are the same age. After 4 billion years, for example, our Sun and all stars with masses smaller than one solar mass will still be on the main sequence while most higher-mass stars will have moved off of the main sequence and become white dwarfs, neutron stars, or black holes.

Astrophysicists can use knowledge of how nuclear fusion works in stars of different masses to determine the surface temperature and luminosity of a star of a given mass at any age. The process works like this: pick a star of a given mass, for example one like the Sun; calculate the surface temperature and luminosity for this star when it is 10 million years old; mark this location on an H-R diagram. The calculations are done using the equations that describe the fundamental physical principles governing gravity, pressure, temperature, the process of nu-

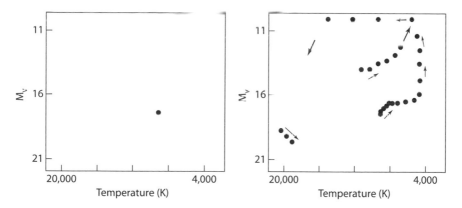

Figure 14.1. Left, a single point marks the temperature and absolute magnitude for a newborn star. Right, the progression of points reveals the changing surface temperature and absolute luminosity of this star as it ages. First it leaves the main sequence and climbs up the H-R diagram as a subgiant and then a red giant; then it drops down onto the horizontal branch and climbs back up to become a giant again; finally it puffs off its envelope as a planetary nebula and leaves behind a bare white dwarf that cools off and fades away.

clear fusion, the processes by which heat is transferred from the core to the surface of the star and radiated out into space, and the elemental composition of the star. Now, imagine that the star shines and burns nuclear fuel for another 10 million years. The elemental composition of the inside of the star will have changed just a little, as helium builds up in the core at the expense of hydrogen. Again calculate the surface temperature and luminosity of the star and again mark the location of the star on the H-R diagram. Repeat these steps at 10-million-year intervals, or in any time increment you prefer. The only limiting factor is the computing power available to you.

Using this method, we would be able to trace the life cycle path on the H-R diagram for a star of any given mass, as its surface temperature and luminosity change with time. For the Sun, the first seven hundred or more points on the diagram, recording about 7 billion years of the star's life cycle in 10-million-year steps, would all fall almost on top of each other, for the Sun remains very stable in temperature and luminosity while it is on the main sequence. Thereafter, the points would trace the Sun's climb onto the subgiant branch and then its rise up the red giant branch. If I were to draw a line that connected all the points as the Sun ages, I would be tracing the life cycle path for the Sun on the H-R diagram. In contrast to the Sun and stars like it, a supermassive, superluminous star might use up all its core hydrogen fuel in its first 10 million years and might become a red giant by the age of 20 million, after only two steps in my calculations.

An even more useful approach is to calculate what happens to an entire cluster containing stars of a wide range of masses. We could mark points on our H-R diagram for stars of every possible mass, running from 0.08 to 50 solar masses, when these stars are all one million years old. If we connected these dots, we would call that line the one-million-year *isochrone*. This isochrone would look like a main sequence, running from the very bottom right to the very top left of our H-R diagram. Next, we could calculate the temperatures and luminosities of all of these stars at an age of 2 million years; if we then marked the locations of these stars on our H-R diagram at 2 million years of age and connected these points, our line would delineate the 2-million-year isochrone. This isochrone would fall almost exactly on top of the

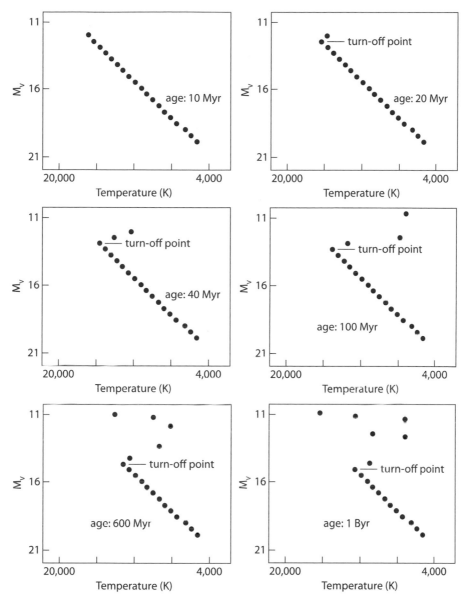

Figure 14.2. Sequence of cartoon snapshots of a star cluster at different ages. Initially (top left), stars of all masses are on the main sequence. After 20 million years (top right), the most massive star has exhausted the hydrogen in its core and has become a subgiant. The most luminous, most massive star still on the main sequence marks the turn-off point and age-dates the star cluster. In the lower panels, less massive stars begin to die and leave the main sequence, while the more massive stars progress to become red giants, planetary nebulae, and, finally, white dwarfs. In any single panel, a line connecting all the dots is called an *isochrone*.

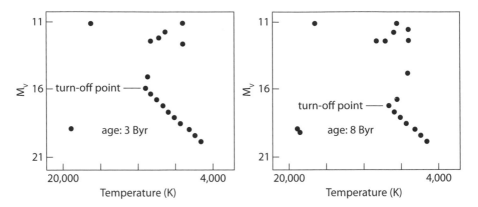

Figure 14.2. (*Continued*)

one-million-year isochrone, since after only two million years all of the stars in this cluster would still be on the main sequence. We could continue this procedure to watch how the entire cluster ages.

At an age of only a few million years, we would notice something happening to the hottest, most luminous, most massive stars: they would have moved off the main sequence and become red giants. An isochrone at one hundred million years would begin at the bottom of the main sequence, extend upwards through most of the main sequence, and then turn off toward the red giants. The most massive stars would by now be red supergiants; the slightly less massive stars would be on the subgiant branch; even less massive stars would still be climbing the red giant branch; and the least massive stars would still be on the main sequence. At 600 million years, even more stars would have migrated off the main sequence. Over the vast eons of time, we would watch the main sequence shrink, from the top left toward the bottom right.

The Turn-off Point

The particular location on the main sequence at which stars turn off the main sequence toward the giant branch is known as the *turn-off point*. The turn-off point determines the age of the cluster. All we need to do is measure its location. For a young cluster, the turn-off point is very high on the main sequence (high luminosity, high temperature). As the stars in the cluster age, the turn-off point marches down the main sequence toward lower luminosities and lower temperatures.

The turn-off point consequently provides a tool for age-dating star clusters. First an astrophysicist makes careful observations of the stars in a cluster in order to place as many of them as possible on an H-R diagram. Then he compares this diagram with theoretical isochrones to find the isochrone that best fits the data. This best-fitting isochrone yields the age of the cluster. Open clusters like the Pleiades (~100 million years) or the Hyades (~500 million years) are young compared to the age of the galaxy and the universe. Globular clusters tend to be of much greater age. The cluster 47 Tuc is ~12 billion years old, while M55 clocks in at 12.5 billion years of age.

Computations of theoretical isochrones are complicated and depend on several important properties of the stars in a cluster, some of which may be very difficult to measure to a high degree of accuracy through observations. For example, the precise elemental composition of a star (how much of the star is composed of hydrogen? of helium? of iron?) will affect a star's temperature and luminosity. Imagine a star with the mass of the Sun made entirely of krypton. This star could not generate energy from proton-proton chain reactions, from the triple-alpha process, or from any of the other nuclear fusion processes available to stars. At birth, it would simply become a white dwarf, albeit an unusual one. To our knowledge, no stars are composed entirely of krypton, and all stars have, very nearly, the same elemental makeup.

In a star like the Sun, about 71 percent of the mass is in the form of hydrogen atoms, 27 percent in the form of helium atoms, and 2 percent in the form of all the other elements (which astronomers refer to, very oddly, as "metals"). Because massive stars fuse hydrogen into helium and helium into heavier elements and then expel most of those fusion products back into space when they die, over billions of years the gas in the Milky Way Galaxy has been enriched in heavier elements at the expense of hydrogen. We expect, as a result, that the youngest stars in the galaxy will have, by proportion, less hydrogen and more helium, carbon, and oxygen than older stars. Conversely

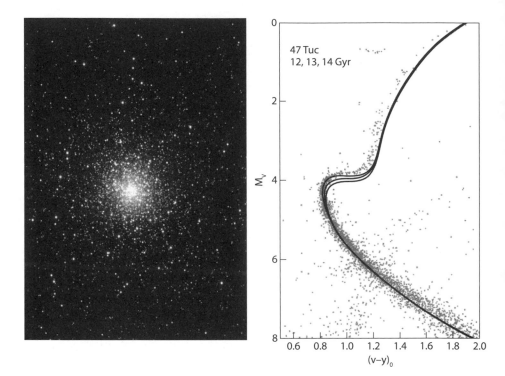

Figure 14.3. Left, image of globular cluster 47 Tuc. Right, H-R diagram (actually a color-magnitude diagram, where color is a proxy for temperature) of the stars in 47 Tuc, showing the main sequence, the turn-off point, the red giant branch, and the horizontal branch. The three isochrones delineate ages of 12 billion years (highest line, near $M_V = 4$), 13 billion years (middle line), and 14 billion years (lowest line). An age of about 12 billion years is the best fit. Atlas Image courtesy of 2MASS/ UMass/IPAC-Caltech/NASA/NSF (left). From Grundahl et al., *Astronomy & Astrophysics* (2002): 395, 481 (right), reproduced with permission © ESO.

(Chapter Twenty-four), the oldest stars in the galaxy, formed when the giant interstellar clouds contained almost no atoms other than those of hydrogen and helium, should be made almost exclusively of hydrogen (~75 percent) and helium (~25 percent), with only very minor amounts of metals. The exact proportion of hydrogen to helium and the exact amount (e.g., 0.0001 percent, 0.001 percent, 0.01 percent or 0.1 percent) of metals found in the chemical composition of a star can make a difference of up to several hundreds of millions of years in calculating the age of a globular cluster from an observed turn-off point. Since all calculations for ages of turn-off points in globular clusters yield cluster ages of 10 or 11 or 12 billion years, however, an error of a few hundred million years is an error of no more than a few percent.

The Ages of the Oldest Globular Clusters

In 2009, the team conducting the ACS (Advanced Camera for Surveys) Survey of Galactic Globular Clusters using the Hubble Space Telescope identified forty-one old globular clusters whose average age is 12.8 billion years. They suggested that the epoch during which the old globular clusters were assembled in the Milky Way was "quick," lasting only about 800 million years, with the oldest forming about 13.2 billion years ago and the youngest as recently as about 12.4 billion years ago.

The globular cluster NGC 6397, which lies at a distance of 2.2 kiloparsecs (7,200 light-years) from Earth, contains about 400,000 stars and is the second closest known globular cluster to Earth. How old is NGC 6397? The turn-off point in its H-R diagram lies in the vicinity of 13 to 14 billion years. The stars that would lie on the main sequence at slightly hotter temperatures are no longer there—they have already become red giants. Recent estimates for the age of this cluster calculated using the main-sequence turn-off point method range from 12.7 to 13.9 billion years. NGC 5904 (M5), which is one of the largest known globular clusters and which lies at a distance of 7.5 kiloparsec (24,500 light-years), has a turn-off point and age that is indistinguishable from that of NGC 6397. NGC 6752 is another relatively close globular cluster, checking in at a distance of 4 kiloparsecs (13,000 light-years) from Earth. The turn-off point for NGC 6752 indicates that, like NGC 6397, this cluster is extremely old, the most recent estimate for its age being 13.4 billion years (within the range from 12.3 to 14.5 billion years).

Based on these age estimates, the globular clusters NGC 6397, 5904, and 6752 appear to be among the oldest globular clusters known. But how confidently do we know these ages? The elemental composition of the stars (how much helium do they contain? how much mass do they

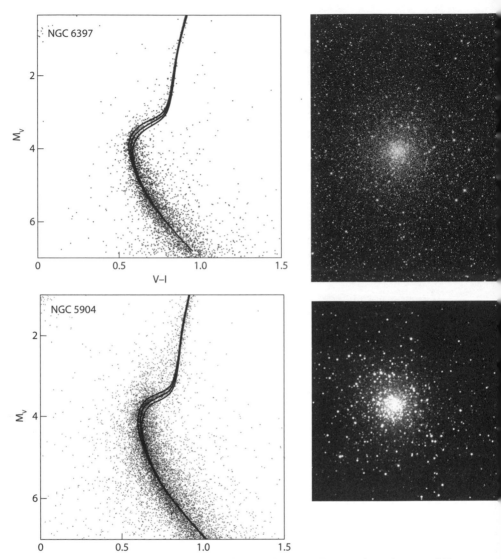

Figure 14.4. Top left, color-magnitude diagram for NGC 6397, with isochrones of 13 (topmost), 14 (middle), and 15 (bottom) billion years overlaid. Top right, Globular Cluster NGC 6397. Bottom left, color-magnitude diagram for NGC 5904, with isochrones of 13 (topmost), 14 (middle), and 15 (bottom) billion years overlaid. Bottom right, Globular Cluster NGC 6904. From Imbriani et al. *Astronomy & Astrophysics* (2004): 420, 625 (top left), reproduced with permission © ESO. Image courtesy of D. Verschatse (Antilhue Observatory, Chile) and ESA/Hubble (top right). From Imbriani et al., *Astronomy & Astrophysics*, (2004): 420, 625 (bottom left), reproduced with permission © ESO. Atlas Image courtesy of 2MASS/UMass/IPAC-Caltech/NASA/NSF (bottom right).

contain in elements heavier than helium?) is not the only difficult-to-measure parameter that could cause errors in our age estimates. Another issue that generates small uncertainties is the limiting accuracy of all astronomical measurements. For example, how well do we know the distances to these clusters? And how much dust (which dims and reddens the light from distant stars) is present in our galaxy between our telescopes and the cluster? We also are limited because our understanding of theoretical physics is imperfect. What are the exact rates at which certain nuclear reactions proceed at a given pressure and temperature? At what rate is heat transported from the insides to the surfaces of stars? In 2003, a team led by Lawrence Krauss, an astrophysicist then at Case Western Reserve University, attempted to quantify all of these uncertainties and concluded that the age of the oldest globular clusters in the Milky Way is between 10.4 and 16.0 billion years. By his reckoning, the age that best fits all the data is 12.6 billion years.

In the end, we cannot use the main-sequence turn-off point method to pin down the ages of globular clusters in our galaxy to the nearest million, 10 million, or even 100 million years. We can, however, say with great certainty that these vast collections of stars are well over 10 billion years old and most likely as much as 13 billion years old. We can also state with high confidence that they are less than 16 billion years old and very likely less than 15 billion years old.

The Age of the Milky Way

Astrophysicists who study the formation of galaxies believe that material from the early universe clumped together to form galaxies, and that objects like globular clusters formed fairly soon thereafter. The ages of globular clusters therefore do not provide the age of the universe or of the Milky Way itself, but instead provide a lower limit to those ages. To determine the age of the Milky Way, we have to answer another question: How much time passed between the moment the Milky Way formed and the epoch when the oldest globular clusters in the Milky Way formed?

We can address this question by recognizing that the material out of which the Milky Way formed should have the elemental composition of the newborn universe: lots of hydrogen, lots of helium, a very little bit of lithium, and nothing else (Chapter Twenty-four). The hydrogen was mostly normal hydrogen, containing a single proton and a single electron, though small amounts of deuterium also formed. The helium was mostly ^{4}He, though smaller amounts of ^{3}He also formed. Almost all of the beryllium that might have formed in the early history of the universe would have been ^{8}Be (four protons, four neutrons), which we know is extremely unstable. None of the ^{8}Be manufactured when the universe was being born could have survived. Stable beryllium, ^{9}Be (four protons, five neutrons), could not have formed during the first moments of the universe; therefore, all of the ^{9}Be in stars had to form later in the history of the universe, in nucleosynthesis reactions that occur when stars explode as supernovae. Any supernovae that existed in the early history of our galaxy must have come from super massive stars with extremely short lifetimes; when these stars died they would have created and injected small amounts of ^{9}Be into the interstellar medium. As each generation of stars succeeded the last, the amount of ^{9}Be in the interstellar medium would have gradually increased. In turn, each generation of newborn stars would have a little bit more ^{9}Be in its makeup than the previous generation.

One way to determine the length of time between the formation of the Milky Way and the formation of the oldest globular clusters in it would be to measure the amount of beryllium in the stars in the oldest such clusters. If they contain any beryllium at all, then some stars must have formed and died before these cluster stars were born. The higher their beryllium content, the more generations of massive stars had to have been born and died before these cluster stars formed. The answer: about one beryllium atom exists for every 2.2 trillion (2.2×10^{12}) hydrogen atoms in the atmospheres of the oldest stars in NGC 6397. Astrophysicists estimate that 200 to 300 million years were necessary before this amount of beryllium accumulated in the Milky Way. Therefore, if the epoch when the oldest known globular clusters formed was about 13 billion years ago, then the moment when the Milky Way Galaxy started forming stars in which light elements like beryllium were synthesized was one-third of a billion years before that, giving the

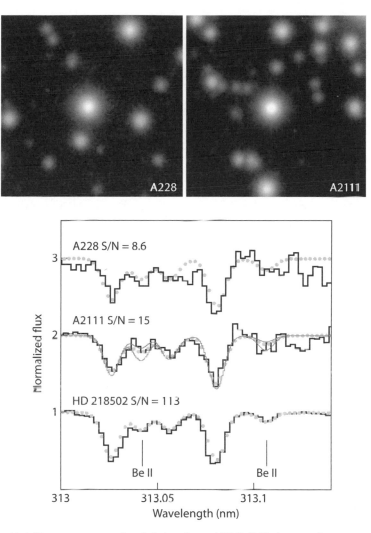

Figure 14.5. Top, two stars in the globular cluster NGC 6397 that are about to age off the main sequence. Bottom, three stars in NGC 6397 showing evidence of the presence of the element beryllium. Image courtesy of ESO (top). From Pasquini et al., *Astronomy & Astrophysics* (2004): 426, 651 (bottom), reproduced with permission © ESO.

Milky Way an age of at least 13.2 or 13.3 billion years. We could fairly conclude that the additional few hundred million years from the birth of the first stars in the Milky Way until the formation of the oldest globular clusters is "in the noise." We know that this approximate amount of time must have elapsed before the first globular clusters appeared, but this knowledge does not help us determine with any greater accuracy the age of the galaxy or of the universe.

Despite some uncertainties in the ages of the oldest globular clusters in the Milky Way, they do provide an independent estimate for the age of the universe that is extremely consistent with age estimates derived from the temperatures and luminosities of the oldest white dwarfs in the galaxy. That consistency adds to our confidence that we are on the right track toward answering our leading question.

• ⠒3⠒ •

THE AGE OF THE UNIVERSE

···CHAPTER 15···

Cepheids

It is worthy of notice that . . . the brighter stars have the longer periods.

—Henrietta Swan Leavitt, in "1777 Variables in the Magellanic Clouds," published in *Annals of Harvard College Observatory* (1908)

In Part II of this book, we followed the path astronomers took in unraveling the astrophysics of stars. We discovered that once we understood how stars generate energy and thus how they live and die, we could, by two independent methods learn the ages of white dwarfs and of globular clusters, some of the oldest objects in the Milky Way and perhaps in the universe. Using these methods we obtained ages for some of the white dwarfs and globular clusters in the Milky Way, which gave us independent but consistent figures for the lowest possible age of the universe (since the universe must be at least a little bit older than the oldest objects in our galaxy). Now, in Part III, we step out further still, beyond the Milky Way, in order to discover methods for making measurements that will give us a more precise age for the universe itself. As a first step into the distant universe, astronomers had to discover that the Milky Way was not the entire universe. Henrietta Swan Leavitt's identification of a number of Cepheid variable stars in two nebulae, known as the Large and Small Magellanic Clouds, provided the vehicle for that monumental discovery.

Leavitt graduated from Radcliffe College, known then as the Society for the Collegiate Instruction of Women, in 1892 and went to work in 1893 as a volunteer computer at the Harvard College Observatory. Soon thereafter, Edward Pickering assigned her the job of identifying

Figure 15.1. Henrietta Swan Leavitt. Image courtesy of AIP
Emilio Segre Visual Archives.

variable stars, these being stars whose output of light varies as a function of time. She quickly became an expert in this task. After three
years of unpaid labor, she delivered a summary report of her findings to
Pickering and departed Cambridge, spending the next two years traveling in Europe and then four more years in Wisconsin as an art in-
structor at Beloit College. Finally, in the summer of 1902, she con-
tacted Pickering and asked permission to return to identifying variable
stars. Clearly pleased with her earlier work, Pickering immediately offered her a full-time, paid position with a wage of thirty cents per hour,
a significant five cents above the standard rate. This decision was one
of the wisest Pickering would ever make. By Labor Day, Leavitt was
again at work in Cambridge, on her way to making one of the most
important discoveries in twentieth-century astronomy, one that laid
the groundwork for another method for measuring the age of the universe, and for the eventual discovery of the expanding universe.

Variable Stars

By 1893, astronomers were already aware that there existed many different kinds of variable stars, though the total number of such stars
that were known to them was small. Tycho Brahe discovered the first

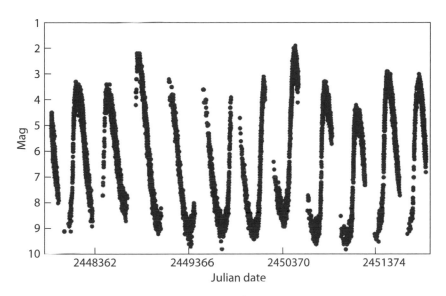

Figure 15.2. A light curve for the variable star Mira, with data from January 1, 1990 through December 31, 2000. The vertical axis numbers measure magnitudes, from one (brightest) to ten (faintest). Mira is known as a long-period variable star, because its light output increases from a minimum brightness of magnitude nine to ten to a maximum brightness of magnitude of two to three and then returns to magnitude nine to ten very regularly over a period of 332 days. Data courtesy of the AAVSO.

variable star when in 1572 he noticed an object he identified as a new star, because previously there had been no star at that location in the sky. This stella nova—we now call the object "Tycho's Supernova"— faded and changed color until, after barely one year, it disappeared from sight altogether. Twenty-four years later, David Fabricius discovered the first periodic variable star, Mira. Mira fades and brightens at a regular and smooth rate, with a period of variability—the length of time needed for the star to fade from maximum brightness to minimum brightness and then to increase again to maximum brightness— of 332 days.

By 1836, 250 years after Tycho identified the first variable star, astronomers had discovered a total of only twenty-six variable stars of any type. The advent of astronomical photography, however, increased the odds that an astronomer comparing photographs of the same patch of sky obtained on multiple nights across days, weeks, or years would notice stellar variability, and indeed by the 1890s, a few hundred such stars had been identified. Leavitt, by herself, would soon be identifying hundreds of variable stars every year, including many called Cepheid variable stars, which proved of primary importance for determining the size, age, and structure of the universe.

Cepheid Variable Stars

The prototype of the variable stars known as Cepheid variable stars, which Henrietta Leavitt would make famous, is Delta Cephei, discovered by John Goodricke in 1784. It was not, however, the first Cepheid discovered. That distinction goes to Eta Aquilae, which was identified as a variable star by Edward Piggot, Goodricke's friend and neighbor, earlier that same year. Eta Aquilae is just over twice as bright at its brightest as at its faintest and has a period of variability of 7.177 days; Delta Cephei changes in brightness by a factor of about 2.3 with a period of variability of 5.366 days. The brightest Cepheid variable star in the sky is Polaris, the North Star, which varies in brightness by only about 3 percent with a period of variability of 3.97 days.

Cepheids get brighter and fainter and brighter again with periods of a few days or weeks, but that is not all they do; they also change color and temperature (and hence spectral type), becoming cooler and redder as they brighten, then warmer and yellower as they fade. In addition, Cepheids differ from other variable stars in the peculiar ways they brighten and fade, peculiarities that depend on their variability periods. For example, shorter period Cepheids (having periods of less than about eight days) brighten much more quickly than they fade. And from the moment at which they are at their faintest, they brighten very steadily, but when they reach maximum brightness and begin to fade, they fade continuously but not steadily. First they fade at an apparently constant rate. But when they are about two-thirds of the way back to minimum brightness, they begin to fade at a slightly slower rate. Then, when they are about 75 percent of the way to minimum light, they

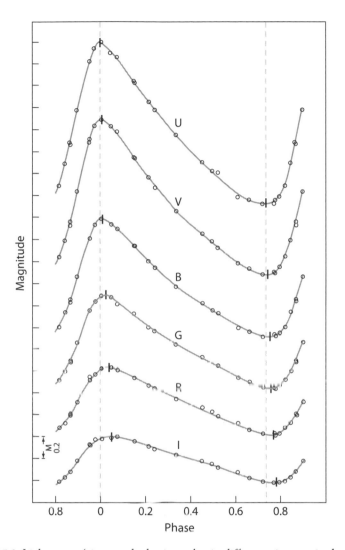

Figure 15.3. Light curve (time on the horizontal axis, difference in magnitude on the vertical axis) for Delta Cephei, as measured at six different colors, from the near ultraviolet (U) to the infrared (I). In visible light (V band), Delta Cephei changes by about 0.9 magnitudes from minimum to maximum (a factor of 2.3 in brightness) over a period of 5.366 days (1.0 units of Phase equals 5.366 days). Unlike the smooth pattern of stars like Mira, Cepheids have a very distinct pattern in their light curves that makes them easy to identify. Notice how Delta Cephei brightens from minimum (Phase = 0.8) to maximum (Phase 0.0) in just over one day but requires an additional four days to again fade to minimum brightness. From Stebbins, *The Astrophysical Journal*, (1945): 101, 47. Reproduced by permission of the AAS.

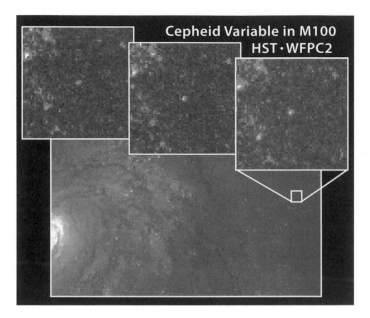

Figure 15.4. The three upper panels show Hubble Space Telescope close-ups of a region in the galaxy M100 in which a Cepheid variable star gets successively brighter (from left to right). Image courtesy of Dr. Wendy L. Freedman, Observatories of the Carnegie Institution of Washington, and NASA.

pick up speed and begin to fade much more quickly again. Despite the quirky nature of this pattern, it is dependable and repeatable. It is the signature, the fingerprints, of a Cepheid with a period of a few days. Cepheids with longer periods also follow characteristic patterns in how they brighten and fade, and these patterns change as the period changes.

Observations from Peru

In 1879, Uriah Boyden, a Boston engineer with no apparent interests in astronomy, left an eccentric bequest: nearly a quarter of a million dollars to any astronomical institution that would build a telescope at a high enough elevation to make more accurate observations than were possible when peering through the Earth's thick lower atmosphere. Ed-

Figure 15.5. Photograph of the Large (left) and Small (right) Magellanic Clouds. Image courtesy of William Keel, University of Alabama at Tuscaloosa.

ward Pickering won the Boyden funds for Harvard in 1887 and, in 1891, used them to establish an observatory at an elevation of 8,000 feet at Arequipa, Peru. Within two years, staff astronomers at Arequipa were obtaining photographic plates of the southern skies and shipping them to Boston. Solon Bailey, the director of the observatory, began a long-term study of one of the greatest globular clusters, Omega Centauri, and by 1901 had discovered 132 variable stars in Omega Centauri.

Among the many objects regularly photographed at Arequipa were two groups of stars visible only from southern latitudes. They were first described by Antonio Pigafetta, a navigator and diarist who sailed with Ferdinand Magellan on his 1519 round-the-world voyage to the Indies, as Nubecula (Nebula) Minor and Nubecula Major and are known today as the Small Magellanic Cloud and the Large Magellanic Cloud, both of which are satellite galaxies of our own Milky Way (although in 1900 neither was known as a galaxy).

Early in 1904, Leavitt discovered several variable stars in a set of photographic plates of the Small Magellanic Cloud. Later that year, she found dozens more in both the Small and the Large Magellanic Clouds. Her discovery rate rose to hundreds per year and eventually she would identify 2,400 such stars. In 1908, Leavitt published, under her own name, "1777 Variables in the Magellanic Clouds" in the *Annals of Harvard College Observatory*. For all of these stars, she was able to determine "the brightest and faintest magnitudes as yet observed," but for sixteen, which she identified in Table VI of her paper, she also was able to determine their periods of variability. "The majority of the

ANNALS OF HARVARD COLLEGE OBSERVATORY. VOL. LX. NO. IV.

1777 VARIABLES IN THE MAGELLANIC CLOUDS.

BY HENRIETTA S. LEAVITT.

Figure 15.6. Henrietta Leavitt was able to publish her 1908 paper in the *Annals of Harvard College Observatory* under her own name. Her follow-up paper would be published in 1912 under the sole authorship of Edward Pickering. From Leavitt, *Annals of Harvard College Observatory* (1908): 60, 87.

light curves [for these variables] have a striking resemblance, in form, to those of cluster variables," Leavitt wrote. That is, they were Cepheids, although they were not yet identified by that name. As for these sixteen stars, she continued, "It is worthy of notice that in Table VI the brighter stars have the longer periods." With historical hindsight, we can recognize this as one of the most understated and important sentences in all of astronomical literature.

The Period-Luminosity Diagram

Four years later, Leavitt would conclude her work on the variable stars in the Small Magellanic Cloud with a brief, three-page paper, "Periods of 25 Variable Stars in the Small Magellanic Cloud," published as a *Harvard College Observatory Circular* under the name of Edward Pickering—though Pickering's first sentence acknowledges that "the following statement has been prepared by Miss Leavitt." Leavitt focused her attention on the sixteen variable stars specifically identified in 1908, along with nine newly identified ones, all of which "resemble the variables found in globular clusters, diminishing slowly in brightness, remaining near minimum for the greater part of the time, and increasing very rapidly to a brief maximum." These are the Cepheids, and those identified by Leavitt had periods that ranged from 1.25 days to 127

TABLE VI.

PERIODS OF VARIABLES IN THE SMALL MAGELLANIC CLOUD.

Harvard No.	Max.	Min.	Range.	Epoch.	Period.	Min. to Max.	Average Dev.	Earliest Observation.	No. Periods.	No. Plates.
					d.	d.				
818	13.6	14.7	1.1	4.0	10.336	1.7	.12	1890	566	44
821	11.2	12.1	0.9	97.	127.	49.	.06	1890	45	89
823	12.2	14.1	1.9	2.9	31.94	3.	.13	1890	184	56
824	11.4	12.8	1.4	.12	65.8	7.	.12	1889	94	83
827	13.4	14.3	0.9	11.6	13.47	6.	.11	1890	448	60
842	14.6	16.1	1.5	2.61	4.2897	0.6	.06	1896	843	26
1374	13.9	15.2	1.3	6.0	8.397	2.	.10	1893	574	42
1400	14.1	14.8	0.7	4.0	6.650	1.	.11	1893	724	42
1425	14.3	15.3	1.0	2.8	4.547	0.8	.09	1893	1042	33
1436	14.8	16.4	1.6	0.02	1.6637	0.3	.10	1893	2859	22
1446	14.8	16.4	1.6	1.38	1.7620	0.3	.09	1896	2052	21
1505	14.8	16.1	1.3	0.02	1.25336	0.2	.10	1896	2335	25
1506	15.1	16.3	1.2	1.08	1.87502	0.3	.09	1896	1560	23
1646	14.4	15.4	1.0	4.30	5.311	0.7	.06	1896	681	24
1649	14.3	15.2	0.9	5.05	5.323	0.7	.10	1893	894	32
1742	14.3	15.5	1.2	0.95	4.9866	0.7	.07	1893	954	28

Figure 15.7. Table VI from Leavitt's 1908 paper, which shows the first evidence for the period-luminosity relationship, contains the key information that enabled astronomers to discover the nature of the spiral nebulae and then to step out into the distant universe and, ultimately, to discover the expanding universe and calculate its age. Of the sixteen stars listed according to their Harvard number (column 1), those with the shortest periods (column 6) are the faintest (the largest maximum and minimum magnitudes; columns 2 and 3), while those with the longest periods are brightest (the smallest maximum and minimum magnitudes). Star 821, for example, has a very long period (127 days) and is very bright (maximum brightness of magnitude 11.2), while star 1505 has a very short period (1.25336 days) and is much fainter (maximum brightness of magnitude 14.8). From Leavitt, Annals of Harvard College Observatory (1908): 60, 87.

days. She then notes, with characteristic restraint, "A remarkable relation between the brightness of these variables and the length of their periods will be noticed . . . the brighter variables have the longer periods." That is, brighter stars blink slowly, fainter stars more rapidly. The graphed version of this relationship is known to astronomers as the *period-luminosity diagram.* Leavitt's keen insight, which is what makes this discovery so important. comes next: "Since the variable stars are probably at nearly the same distance from the Earth, their periods are apparently associated with their actual emission of light."

In 1912 astronomers did not know the distance to the Small Magellanic Cloud; however, since it was sort of an agglomeration of stars, it

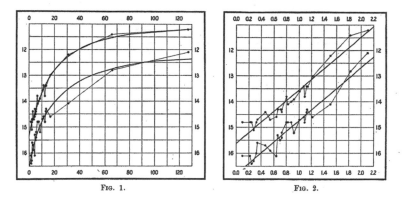

Fig. 1.   Fig. 2.

Figure 15.8. Henrietta Leavitt's plots of the periods of "cluster variables" in the Small Magellanic Cloud (x-axis) versus the brightnesses of those stars (y-axis). Each plot shows two lines, the top line indicating the maximum brightness (minimum apparent magnitude) of the stars and the bottom line the minimum brightness (largest apparent magnitude) of the stars. The left plot marks the period of variability for these stars in days, from 0 to 140 days (x-axis); the right plot marks the period using the logarithm of the number of days (x-axis). These plots reveal the simple relationship between the magnitude and period of Cepheid variable stars that has become known as the *period-luminosity relationship*. From Pickering, *Harvard College Circular* (1912): 173.

was clear that, as in any other cluster, the distances between the variable stars must be negligible compared to the distance to the Small Magellanic Cloud itself. As a result, even though Leavitt had no idea what the intrinsic luminosities were for her twenty-five stars, she could make direct comparisons of their apparent luminosities and say, with absolute certainty, that the *differences* in the apparent magnitudes of these stars were identical to the *differences* in their absolute magnitudes. The brighter stars were truly brighter, the fainter stars truly fainter. She could say, again with absolute certainty, that the period of time over which a given cluster variable (Cepheid) star varied from maximum to minimum and back again to maximum light output *determined the absolute magnitude of that star.*

What does Leavitt's discovery mean? If we can measure the period of variability for any single Cepheid variable star, we instantly know the absolute magnitude of that star. Since we can directly measure the apparent magnitude of the Cepheid, the combination of the period (which gives the absolute magnitude) and the apparent magnitude yields the distance to the star. This is an incredibly powerful discovery. In 1912, however, there was one important catch: Leavitt did not know the distance to the Small Magellanic Cloud, so she did not know the absolute magnitude for even a single Cepheid. Her period-luminosity diagram was not calibrated.

This situation is akin to knowing that you have a fleet of different model cars, each with a unique capacity gas tank, but all with the same mileage rating (miles per gallon). Only if you know what that mileage rating is can you could calculate the total driving distance possible on a single tank of gas for each and every car. For Cepheids, the period-luminosity relationship enables us to calculate the distance to any Cepheid (or object containing a Cepheid), *provided* that we have first calibrated the period-luminosity diagram by measuring the distance to, and thereby the absolute magnitude of, just one Cepheid.

Leavitt was expert at identifying variable stars, but her job duties were narrowly defined and did not extend to designing new observation projects to determine the distance to one or more cluster variables. She could only write, as she did in "her" 1912 paper with characteristic forthrightness and modesty, "It is to be hoped, also, that the parallaxes to some variables of this type may be measured." The work of calibrating the period-luminosity relationship would be done by others.

Cluster-type Variables

During the same decade when Leavitt was discovering Cepheid variable stars in the Small Magellanic Cloud, Solon Bailey, helped greatly by another of Pickering's female computers, Evelyn Leland, discovered another kind of variable star in globular clusters in the Milky Way. These variable stars, which were so strongly associated with globular clusters that Bailey identified them as *cluster-type variable stars*, were those to which Leavitt had referred when she said in 1908 that "[t]he majority of the light curves [for the Cepheid variables] have a striking resemblance, in form, to those of cluster variables."

The big difference between Cepheids and cluster-type variables was the period. Cepheids had periods ranging from days to months, while cluster-type variables had periods of hours. By 1913, Bailey could say with confidence, based on data from the 110 variables whose periods and brightnesses had been most securely determined, that cluster-type variables all had periods between about four hours and one day, and "are all of about the same brightness." Notably, Bailey also found a handful of Cepheid-type variables in globular clusters.

Calibrating the Period-Luminosity Relationship

It was none other than Ejnar Hertzsprung who calculated distances to a handful of Cepheids, and he did so almost immediately after Leavitt had discovered the period-luminosity relationship. Though not a single cluster variable star was close enough to make a traditional parallax measurement, astronomers had developed a tool known as *statistical parallax* that Hertzsprung applied to a total of thirteen stars, including Polaris and Delta Cephei, that he thought were similar to Delta Cephei and to Leavitt's variables in the Small Magellanic Cloud. He identified these stars as Delta Cephei stars, thereby giving them the name we have used for them ever since. On the basis of the results obtained from his statistical parallax computations, he next calculated the distance to the Small Magellanic Cloud and thereby calibrated Leavitt's period-luminosity relationship.

The method of statistical parallax takes advantage of the fact that the Sun travels at a speed of about twenty kilometers per second relative to the average motion of very nearby stars, and as it travels it drags the Earth along with it through space. Together, the location of the Earth-Sun system changes by about four astronomical units (AU), annually. As a result, the baseline from which Earth-bound observers measure the motions of stars increases by about 40 AU per decade. Over many years, because of the changing position of the Sun and Earth in the galaxy, the directions in which stars appear to move will change; the directions in which the stars are actually moving of course do not change. But the apparent change in the direction of motion of stars due to the real motion of the Sun and Earth will be greater for closer stars, smaller for more distant stars.

Hertzsprung used this method to calculate distances to nearby Cepheids, and he found that the absolute magnitude of Cepheids with periods of 6.6 days was M = −2.3, which is about 7.0 magnitudes (630 times) brighter than the Sun in visible light. When he then applied this absolute brightness calibration to Leavitt's Cepheids in the period-luminosity relationship, he found that the Cepheids in the Small Magellanic Cloud lay at a distance that corresponds to a parallax of p = 0.0001 seconds of arc, or 10 kiloparsecs (33,000 light-years). Amazingly, in 1913 the most distant celestial object for which a distance was securely determined was the Hyades Cluster, then thought to be at a distance of about 40 parsecs (130 light-years). Hertzsprung had just used Leavitt's period-luminosity diagram and his measurements of thirteen variable stars in the Milky Way to reach about 250 times further out into the universe than any astronomer had done before.

We now know that Hertzsprung underestimated the absolute brightnesses of Leavitt's Cepheids by a factor of about 40 and consequently underestimated the distance to the Small Magellanic Cloud by a factor of about six. In 1917, Harlow Shapley (we will hear a good bit more about him in the next chapter), working at Mount Wilson Observatory in California, redid Hertzsprung's trigonometric parallax calibration. Shapley, however, used only eleven of Hertzsprung's thirteen stars, as he was convinced that the light curves for two of them were atypical and should not be grouped with the others (in fact he concluded that they were a different kind of variable star and not Cepheids). He found that a Cepheid with a period of 5.96 days would have the absolute magnitude M = −2.35, very nearly the same result as found by Hertzsprung. Shapley then plotted on a single graph the combined periods and apparent luminosities of his eleven nearby Cepheids, Leavitt's Magellanic Cloud Cepheids, and Bailey's cluster-type and Cepheid-like variables in globular clusters and used his absolute magnitude calibration for the 5.96-day Cepheid to generate a calibrated period-luminosity diagram.

From this initial calibration, Shapley calculated more accurate values—underestimated in comparison to modern measurements by

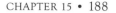

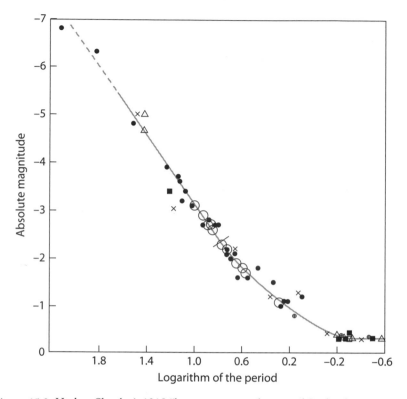

Figure 15.9. Harlow Shapley's 1918 "luminosity-period curve of Cepheid variation." This plot shows the logarithm of the period (negative numbers are periods less than one day; 0.0 is a period of 1 day; +1.0 is a period of 10 days; +2.0 would be a period of 100 days) along the x-axis and the (median) absolute magnitude for each star on the y-axis (0 is faintest; –7 is brightest). The Cepheids with the longest periods are brightest, those with the shortest periods are faintest. From Shapley, *The Astrophysical Journal* (1918): 48, 89.

only a factor of about 10—for the brightnesses of Leavitt's Cepheids and derived a distance to the Small Magellanic Cloud of 19 kiloparsecs (60,000 light-years; *p* = 0.000053 seconds of arc). Three more decades would pass before astronomers would be able to calibrate the brightnesses of Leavitt's Cepheids correctly to within a few percent, but Shapley's imperfect results nonetheless represented significant progress toward an accurate measurement of the distances to the Magellanic Clouds.

Over the next thirty years, many astronomers confirmed Shapley's work and, as it turned out, all of them were wrong. Only in the 1950s (Chapter Twenty) did astronomers discover that the stars on the period-luminosity diagram, as calibrated by Shapley and many others in the 1920s, 30s, and 40s, included two different types of Cepheid variable stars, now known as Type I (brighter) and Type II (fainter) Cepheid variables, and that the short-period cluster variable stars (first identified by Bailey and Leland) are not Cepheids (they are now known as *RR Lyrae variables*). The Cepheids found by Leavitt in the Small Magellanic Cloud are Type I Cepheids; the nearby Cepheids used by Shapley for his calibration work are Type IIs. The two types have intrinsic brightnesses that differ by a factor of four (1.5 magnitudes) and therefore describe two different lines, not a single line, on a period-luminosity diagram. By using the fainter Type II Cepheids to calibrate (incorrectly) the brightnesses of the brighter Type I Cepheids, Shapley mixed apples with oranges. He underestimated the absolute luminosities of Leavitt's Type I Cepheids by about a factor of four and therefore underestimated their distances by a bit more than a factor of two. When, in the 1950s, astronomers were able to determine correct intrinsic brightnesses for the Cepheids in the Small Magellanic Cloud, the measured distance to the Small Magellanic Cloud more than doubled.

Because it is so important to their work, astronomers continue to try to improve the accuracy with which they calibrate the period-luminosity relationship. One century after Leavitt discovered it and Hertzsprung

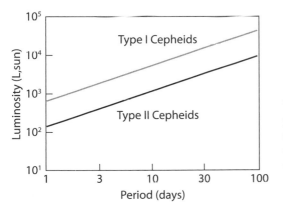

Figure 15.10. A modern period-luminosity diagram showing the two types of Cepheids. Type I Cepheids are about four times brighter than Type II Cepheids having the same period.

and Shapley first attempted to calibrate it, astronomers are now confident that the calibrated values for the relationship are accurate to within a few percent. With this best modern calibration, Michael Feast and
Robin Catchpole in 1997 calculated the distance to the Large Magellanic Cloud as 55 kiloparsecs (180,000 light-years) and the distance to
the Small Magellanic Cloud as 64 kiloparsecs (210,000 light-years). In
the end, Hertzsprung was wrong by a factor of six and Shapley was
wrong by a factor of three; both, though, were more right than they
were wrong, as both had correctly determined the general length scale
(tens of kiloparsecs, not a few hundred parsecs or many megaparsecs)
for these vast interstellar distances. Thanks to Leavitt's variable stars,
astronomers found themselves on the cusp of a decade of absolutely
monumental discoveries. First, the identification of Cepheids in spiral
nebulae would enable astronomers to measure the distances to these
enigmatic objects, proving once and for all that the spirals are distant
galaxies; then, the discovery that the distances and velocities of these
galaxies are correlated would lead to the discovery of the expanding
universe. Ultimately, the expanding universe measurements would present us with another method for determining the age of the universe.

CHAPTER 16

An Irregular System of Globular Clusters

We have found that the center of the elongated and somewhat irregular system of globular clusters lies in the plane of the Milky Way . . . we estimate provisionally, the distance of the center to be 13,000 parsecs.

—Harlow Shapley, in "Studies Based on the Colors and Magnitudes in Stellar Clusters. Seventh Paper: The Distances, Distribution in Space, and Dimensions of 69 Globular Clusters," in *The Astrophysical Journal* (1917)

Many great topics of debate in nineteenth-century astronomy were well on their way to resolution with the invention of spectroscopy, the creation of the Harvard spectral classes, and the discovery of the Hertzsprung-Russell diagram. But early in the twentieth century, perhaps the greatest of all nineteenth-century puzzles in astronomy remained unsolved: What is the nature of the great *spiral nebulae*? Are they "island universes," what we today would call galaxies, similar to the Milky Way, or are they nebulae embedded within and intrinsically part of the Milky Way. Stated more directly, is the Milky Way the entire universe or is the Milky Way one small galaxy in a universe filled with cousin galaxies?

The Island Universe Debate Begins

The great debate over the nature of spiral nebulae began in 1845 when Irish astronomer William Parsons (Lord Rosse) drew a sketch of one of

Figure 16.1. Top, sketch of the Whirlpool Galaxy (M51) made by Lord Rosse in 1845. Bottom, Hubble Space Telescope image of M51. Image courtesy of NASA, ESA, S. Beckwith (STScI), and the Hubble Heritage team (STScI/AURA).

the nebulous objects in Messier's catalog—namely M51 (the Whirl-pool Galaxy)—which showed clear evidence of a spiral structure. Lord Rosse had, during the years 1842–45, built what was at the time the largest telescope in the world; it had a six-foot diameter mirror and was known as the Leviathan of Parsonstown after its location in Birr Castle near Parsonstown, Ireland. With his technological marvel, Lord Rosse could see details in the heavens never before viewed by any earthbound observer and soon had drawn sketches of the spiral structures he saw in M33 (the Triangulum Galaxy), M99, and M101 (the Pinwheel Gal-axy). Was each spiral nebula a single newborn star still wrapped in its swirling placental cloud and located so close to the Sun that we could see details in the spatial structure of the swirls? Or, as predicted in the eighteenth-century speculations of Immanuel Kant, were spiral neb-ulae distant island universes, each one a disk comprised of countless stars.

The debate raged but did not progress for half a century. It finally lurched forward in 1898, when Julius Scheiner of Potsdam Observatory (20 km southwest of Berlin) obtained, in a 7.5-hour exposure, a spec-trum of the Andromeda Nebula (M31), and found that it looked ex-tremely similar to the spectrum of the Sun. He reasoned that "our stel-lar system" (the Milky Way), if viewed from a great distance, would similarly look like a star and concluded that Andromeda was a group of stars so distant that no individual stars could be distinguished with even the largest telescopes.

Score: Island Universes = 1, Local Star-forming Cloud = 0.

The next important measurement in the spiral nebula debate was made by Vesto Slipher of Lowell Observatory in Arizona, who in 1912 reported in the *Lowell Observatory Bulletin* (No. 55) that he had ob-tained a twenty-one-hour exposure (over three nights) of the spectrum of "the nebula in the Pleiades." (Known as NGC 1435 or "Tempel's Nebula," it is a faint, diffuse nebula about the size of the full Moon that surrounds the star Merope.) Slipher noted that the spectrum "contains no traces of any of the bright lines found in the spectra of gaseous neb-ulae" (such as Orion) and that "the whole spectrum is a true copy of that of the brightest stars in the Pleiades." He speculated that the light from this nebula could be the collected light from many, very distant stars; but he wondered why the sum of the light from so many stars appeared as a single spectral type. Shouldn't the summed spectrum from the light of many stars appear as a blend of many spectral types? He concluded, correctly, that the light in this spectrum must therefore be from a single star reflecting off cloudy matter near the star. This result, he concluded incorrectly, "suggested to me that the Andromeda Nebula and similar spiral nebulae might consist of a central star enveloped and beclouded by fragmentary and disintegrated matter which shines by light supplied by the central sun."

Score: Island Universes = 1, Local Star-forming Cloud = 1.

In 1913, Slipher reported, in another *Lowell Observatory Bulletin*, that he had measured the radial velocity of Andromeda and determined that it was moving toward the Sun with the highest blueshifted velocity (300 kilometers per second) ever recorded for any object. Perhaps, he concluded, the Andromeda Nebula had passed very close to a "dark star" and, presumably as a result of that close encounter, been propelled at high speed toward the solar system.

Score: Island Universes = 1, Local Star-forming Cloud = 2.

Inspired by the "quite exceptional velocity" he had measured for Andromeda, Slipher expanded his work to include radial velocity measurements of other spiral nebulae, all of which, he concluded, have spectra similar to that of Andromeda in the sense that each looks like a spectrum from a single star and not like a composite of spectra from stars of many spectral types. By 1915, he had measured the radial velocities of ten other spiral nebulae. One of these, NGC 221 (also known as M32), which appears in the sky as a very near neighbor of Andromeda, had the same blueshifted velocity as Andromeda. The other nine spirals all had redshifted velocities ranging from 200 kilometers per second to 1,100 kilometers per second; the average velocity of all eleven spiral nebulae was 400 kilometers per second, though the average velocity *away* from the Earth for the nine nebulae showing redshifts was 550 kilometers per second and three of these spirals had velocities of either 1,000 or 1,100 kilometers per second. As the spirals had an average velocity at least twenty-five times greater than the average stellar velocity, Slipher decided that the spirals were probably very evolved stars, as such high velocities were consistent with a then-prevalent theory that stellar velocities increase with stellar spectral type and, implicitly, with age.

Score: Island Universes = 1, Local Star-forming Cloud = 3.

Distances to Globular Clusters and the Shape of the Milky Way

Harlow Shapley completed his Ph.D. in 1914, studying eclipsing binary stars under the great Henry Norris Russell. After finishing his graduate work at Princeton, he took a job at Mount Wilson Observatory near Pasadena, California, home of the then-largest telescope in the world, the 60-inch reflector completed in 1908. (The main light-collecting focusing device in such a telescope is a shallow bowl-shaped reflecting mirror with a diameter of 60 inches. A few years later, in 1917, Mount Wilson, would become home to the behemoth 100-inch telescope that would remain the biggest in the world until 1948, when it was surpassed by the 200-inch Palomar Mountain telescope, located just north of San Diego. Shapley was in the right place at the right time to do extraordinary work, and he then proceeded to do just that.

During his first few years at Mount Wilson, Shapley worked to understand Cepheids and to identify more of them in star clusters. Astronomers had quickly come to recognize that the brightness of a Cepheid varied because of some intrinsic properties of the star, probably due to pulsation, and not due to external processes, such as eclipses. They were recognized as intrinsically bright stars, so bright that they could be observed at great distances, especially with big telescopes. Furthermore, they were fairly easy to identify—if you were willing to spend many long nights peering through a telescope—because of their unique pattern of variability. A serious search for distant Cepheids was an ideal project for a highly skilled and motivated young astronomer who had access to large amounts of observing time on Earth's largest telescope.

In the course of his research, Shapley pondered whether globular clusters might also be island universes located far beyond the confines of the Milky Way. Or, as was the growing consensus with regard to spiral nebulae, were they objects located within the Milky Way? Since Cepheids could be found in globular clusters, and since the period-luminosity diagram could be used to determine the distance to any celestial object containing a Cepheid, Shapley took advantage of his position as an astronomer at Mount Wilson to search for Cepheids in globular clusters, hoping to use them to determine the distances to these clusters and thereby to definitively resolve the island universe debate.

Through his studies of the period-luminosity relationship for Cepheids, Shapley was able to calculate not only the distance to the Small Magellanic Cloud (measured by him at 19 kiloparsecs), but also to seven globular clusters (his measured distances ranged from 7 to 15 kiloparsecs). In parallel, he also developed two other tools for measuring distances to globular clusters, both of which were based on the assumption that all globular clusters are sufficiently similar to permit us to think of them as identical:

- If globular clusters are identical groups of 100,000 or more stars, each confined to a volume of space only a few tens of parsecs across, then if we saw every globular cluster at the same distance, each would subtend the same angular size in the sky. Therefore, more distant clusters would appear smaller; closer clusters would appear larger.
- If all stars in a single globular cluster are at about the same distance and each has an almost uncountable number of stars, we should expect that the average brightness of the twenty-five brightest stars in one cluster would be the same as the average brightness of the twenty-five brightest stars in any other cluster. Therefore, in more distant clusters, the average brightness of the twenty-five brightest stars would appear fainter than the average brightness of the twenty-five brightest stars in more nearby clusters.

From these three independent measures—the average brightness of the Cepheids in each cluster (i.e., the period-luminosity relationship), the angular sizes of the clusters, and the average brightness of the brightest twenty-five stars in each cluster—Shapley calculated distances for

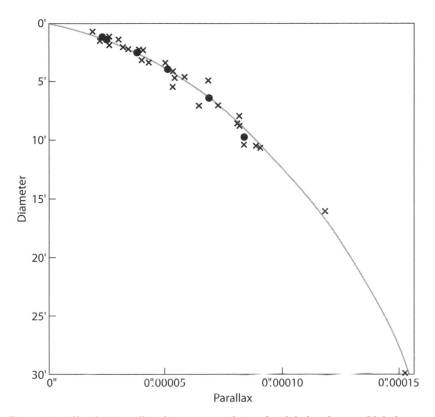

Figure 16.2. Shapley's parallax-diameter correlation for globular clusters. Globular clusters that subtend smaller angular sizes in the sky (higher on graph) have smaller parallaxes (left on graph) and therefore lie at greater distances than the globular clusters with larger angular sizes. From Shapley, *The Astrophysical Journal* (1918): 48, 154.

each of the seven globular clusters for which he was able to apply the period-luminosity relationship. In doing so, he found a remarkable consistency between the distances derived from the period-luminosity diagram and the distances derived from the other two measurements. This consistency demonstrated that each of the three methods could independently be used to determine the distance to a globular cluster, yet only one of the three depended on identifying Cepheids in any single globular cluster. He therefore could use the two new techniques to measure distances to globular clusters without Cepheids, of which there were many.

Shapley then applied his new techniques to determine distances to sixty-two additional globular clusters in which no Cepheids had yet been identified. For twenty-one of these, he could apply both his magnitude and angular diameter methods. For the other forty-one, he could employ only his angular diameter method. Ultimately, Shapley determined distances to all sixty-nine star clusters known in 1917 to be globular clusters. He then had three pieces of information for each of them: latitude and longitude (position in the sky) and distance. He used these measured quantities to examine the three-dimensional distribution of globular clusters and found that the Sun was in an "eccentric position;" that is, the Sun, quite unexpectedly, was located far from the center of the overall distribution of globular clusters. In fact, in a plot locating these clusters on the sky, sixty-four of the sixty-nine were in one half of the sky; the Sun was located on the imaginary line that divided the sky into halves; and the other five clusters were just barely across the dividing line in the other half of the sky. With respect to the plane of the Milky Way, the globular clusters were distributed equally above and below the plane, but none were within 1,300 parsecs (400 light-years) of the mid-plane of the Milky Way. Shapley wrote: "We may say confidently that the plane of the Milky Way is also a symmetrical plane in the great system of globular clusters. . . . We have found that the center of the elongated and somewhat irregular system of globular clusters lies in the plane of the Milky Way." In terms of distance, the overall distribution of globular clusters extended from the Sun out to a distance of about 67 kiloparsecs, with the center of the distribution located in the direction of the constellation Sagittarius. "We estimate provisionally," he concluded, "the distance of the center to be 13,000 parsecs." This was a tour de force. On a December day in 1917, Harlow Shapley had shifted the center of the universe away from our Sun, the place where Copernicus had fixed it in 1543. Shapley moved it to a spot more than 40,000 light-years away, deep in the heart of the Milky Way.

As measured by Shapley, the distance to the Small Magellanic Cloud of 19 kiloparsecs was three times closer than the distance to the most distant globular cluster known in the Milky Way. Shapley's work there-

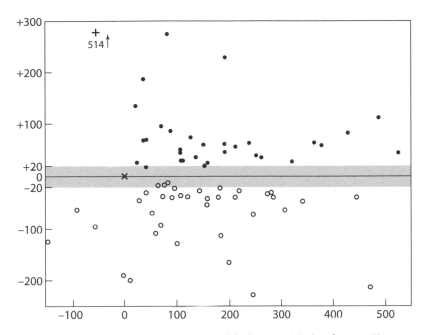

Figure 16.3. Shapley's map of the positions of the known globular clusters. The vertical axis marks distance above the midplane of the Milky Way, while the horizontal axis marks the distance from the Sun. All units are in increments of 100 parsecs, so that one square (100 units) equals a distance of 10,000 parsecs. The Sun is at the position x = 0 and y = 0. Shapley's map shows that all but five of the globular clusters are on one side of the sky, placing the Sun near the edge of this distribution. The clusters are also distributed equally above and below the galactic midplane. From Shapley, *The Astrophysical Journal* (1918): 48, 154.

fore appeared clearly to affirm that the Magellanic Clouds and the spiral nebulae like Andromeda must be parts of the Milky Way.

   Score: Island Universes = 1, Local Star-forming Cloud = 4.

Spinning Spirals Must Be Nearby

The Dutch-American astronomer Adrian van Maanen, also working at Mount Wilson Observatory, put the nail in the coffin of the island universe theory. Or so it seemed. His observations revealed the appar-

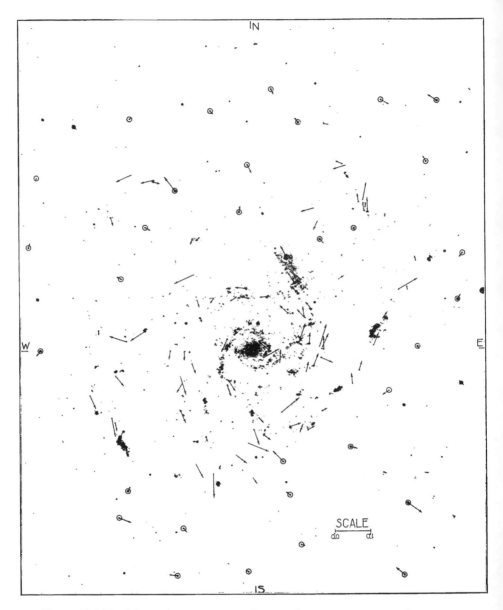

Figure 16.4. Van Maanen's measurements showing the apparent rotation of stars in M101. The arrows indicate the direction and magnitude of the (incorrectly) measured annual motions. From van Maanen, *The Astrophysical Journal* (1916): 44, 210.

ent proper motions of stars in several spiral nebulae, and in his published papers with these results, he called attention to the change in position of individual stars along the length and breadth of the spiral arms. These results appeared to provide evidence for the physical movement of material along the spiral arms. Logically, if the spirals are extremely distant, then the spirals themselves must be enormous, given how big they appear in the sky; if they are enormous, then even if the stars are moving at incredibly great speeds, any streaming motion of the stars in those nebulae should be immeasurably small in angular size. On the other hand, if such motions could be measured—and van Maanen claimed that he had measured them—then the spirals must be small and relatively nearby.

In 1916, van Maanen published his results for spiral nebula M101, followed over the next seven years by independent measurements for M33, M51, M63, M81, M94, and NGC 2403, all of which, according to van Maanen, showed strong evidence "in favor of a motion along the streamers of the spiral, combined with some radial outward motion." In 1922, van Maanen wrote that "these methods point to parallaxes of from 0.0001″ to 0.0010″ and indicate that the distances and sizes of the nebulae are considerably less than those required of the island universe theory." These parallaxes, which correspond to distances ranging from 1 to 10 kiloparsecs, place these seven spiral nebulae well within Shapley's Milky Way.

Score: Island Universes = 1, Local Star-Forming Cloud = 11.
Game, set, match.

About seventy-five years after Lord Rosse drew his first sketch of a spiral nebula, the debate appeared to be over. The steady accumulation of evidence from a decade of work by Slipher, Shapley, and van Maanen appeared to have driven the astronomical community firmly toward broad consensus. Spiral nebulae are part of the Milky Way. They are not island universes. The Milky Way encompasses the entire universe.

A few years later, new observations proved this consensus completely wrong.

CHAPTER 17

The Milky Way Demoted

The present investigation identifies N.G.C. 6822 as an isolated system of stars and nebulae of the same type as the Magellanic Clouds, although somewhat smaller and more distant. . . . The distance is the only quantity of a new order.

—Edwin Hubble, in "N.G.C. 6822, A Remote Stellar System," *The Astrophysical Journal* (1925)

In the third decade of the twentieth century, the golden boy Edwin Hubble, standing on the shoulders of Henrietta Leavitt and Harlow Shapley, peered further out into the universe than anyone believed possible. What he saw put an end once and for all to the great debate over the nature of spiral nebulae.

As a high school student, Edwin Hubble was the 1906 Illinois state champion in the high jump, setting a state record in the event. As an undergraduate at the University of Chicago, he was on Big Ten championship teams in both track and basketball before he was awarded a Rhodes Scholarship. As a Rhodes Scholar, he was a star athlete in track, water polo, and boxing while reading Roman and English law at Queen's College, Oxford. In 1913, he returned stateside and taught Spanish and coached basketball at a high school in New Albany, Indiana. After a year of that, he returned to the University of Chicago, where he earned his doctoral degree in astronomy in 1917. The day he defended his doctoral thesis, he enlisted in the army for duty in what was then called the Great War. Commissioned as a captain, he served briefly in France and was quickly promoted to major. After his release from the army in 1919, he accepted a job at Mount Wilson Observa-

Figure 17.1. The Mount Wilson Observatory's 100-inch telescope, c. 1940. Hubble's chair is visible to the left. Image courtesy of the Huntington Library, San Marino, California.

tory, studying spiral nebulae. It was his good fortune that the 100-inch Hooker telescope, the largest and best telescope on Earth, had just been commissioned at Mount Wilson.

The Distance to Andromeda

Hubble's big break came on the 23rd of October, 1923, when he discovered the first Cepheid known in a spiral nebula. He wrote in a letter to Shapley, who by then was director of the Harvard College Observatory, "You will be interested to hear that I have found a Cepheid variable in the Andromeda Nebula . . . I have followed the nebula . . . as closely as the weather permitted and in the last five months have netted nine novae and two variables. . . ." Over the next year, Hubble made good use of his ready access to the Hooker telescope. He obtained 130 photographic plates of Andromeda (the product of 130 nights spent observing the nebula) and 65 photographic plates of a second spiral, M33. With these data in hand, he identified a large number of Cepheids in each spiral and used these and the period-luminosity relationship to determine the distance to Andromeda. His answer: almost 300 kiloparsecs (1 million light-years). If Hubble was right, then Andromeda was far outside of the Milky Way. It was in fact an island universe, and Shapley was wrong about the size of the universe and the nature of the spiral nebulae. Shapley challenged Hubble on the battleground of ideas, arguing that Cepheids with periods longer than about thirty days were unreliable (in Shapley's work, he had almost no stars with such long periods) and that Hubble's distance to Andromeda, which depended in part on data from several long period Cepheids, must be wrong.

In late 1924, Hubble wrote to Shapley: "The straws are all pointing in one direction and it will do no harm to consider the various possibilities involved." Shapley replied, "I do not know whether I am sorry or glad . . . perhaps both." Whether sorry or glad, the die was cast. On January 1, 1925, Henry Norris Russell read a paper submitted by the absent Hubble at a joint meeting of the American Astronomical Society and the American Association for the Advancement of Science in Washington, D.C. Russell presented Hubble's distance measurements for Andromeda and for M33, which showed them both to be at distances of 285 kiloparsecs (930,000 light-years). In his paper on Andromeda and M33, which would appear in April of 1925 in *Popular Astronomy*, Hubble left the dramatic conclusion unstated: both Andromeda and M33 lie far outside of the Milky Way. Hubble was awarded $500 for the most outstanding research paper presented at the meeting and instantly became a major force in the world of astronomy.

Concurrently with his work on Andromeda, Hubble was also observing a fuzzy cloudlike object known as Barnard's Galaxy (NGC 6822), named for its discoverer, E. E. Barnard. Barnard had first sighted the object in 1884, looking through the eyepiece of Vanderbilt University's six-inch telescope. Studying fifty images of NGC 6822, Hubble found fifteen variable stars, including eleven Cepheids with periods

Figure 17.2. Infrared image of NGC 6822. NGC 6822 is the first galaxy identified by Edwin Hubble as an object clearly outside of the Milky Way. Image courtesy of 2MASS/UMass/IPAC-Caltech/NASA/NSF.

ranging from twelve to sixty-four days. The distance to Barnard's Galaxy, derived by Hubble from the period-luminosity relationship was 250 kiloparsecs (700,000 light-years). Hubble's September 1925 paper describing this research noted, fairly casually but unequivocally, that NGC 6822 is "the first object definitely assigned to a region outside the galactic system."

Hubble's universe of 1925 bore little resemblance to Shapley's 1924 universe. Overnight, the Milky Way had been demoted. No longer the universe entire, it was just another among the many spiral galaxies and

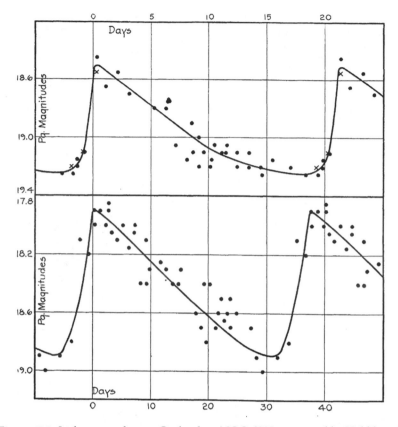

Figure 17.3. Light curves for two Cepheids in NGC 6822, presented by Hubble in his groundbreaking 1925 paper. From Hubble, *The Astrophysical Journal* (1925) 62, 409. Reproduced by permission of the AAS.

other nebulae that filled a vastly expanded universe. The Milky Way had a diameter of 30 to 100 kiloparsecs (100 to 300 thousand light-years). The island universes, now recognized as other galaxies, lay at distances of hundreds of thousands to millions of parsecs.

A Decade of Errors

How had the astronomy community, including Slipher, van Maanen, and Shapley, been led so far astray? Slipher made good measurements, but he misinterpreted them; he naively assumed that all nebulae are alike and that one interpretation would serve for them all. He was wrong. Some nebulae (the nebula in the Pleiades, for example) are small, nearby, wispy clouds reflecting light from a few stars in their vicinity. Others (like Andromeda) are vast and distant collections of hundreds of billions of stars that shine by emitted light. His mistake was avoidable.

Slipher knew that the nebula in Orion had bright lines that did not show up in the nebula in the Pleiades; the Orion Nebula spectrum was very unlike that of a star, while the nebula in the Pleiades was "a true copy of that of the brightest stars in the Pleiades." He had enough information to recognize that not all nebulae are alike, even though he did not have enough information about the spectrum of Andromeda or of any other spiral nebula to know whether they resembled the nebula in the Pleiades or the nebula in Orion or were unique. His conclusion that the "Andromeda Nebula and similar spiral nebulae might consist of a central star enveloped and beclouded by fragmentary and disintegrated matter which shines by light supplied by the central sun" was rash and not supported by sufficient evidence.

Van Maanen's measurements of rotating galaxies with material streaming along their spiral arms were specious, never confirmed, and never reproduced by any other astronomer. Van Maanen made irreproducible measurements that were simply incorrect. Eager to discover what he believed must be true, he misled himself, finding the answers he thought should be in his data rather than the answers truly revealed by his photographs. His mistake was science done poorly. Hubble was, at least publicly, very gracious. In 1935, after reviewing van Maanen's work, he kindly suggested that van Maanen's errors could perhaps be explained by his failure to recognize systematic errors in his own measurement work. Van Maanen nevertheless continued to insist that "the persistence of the positive sign [in the streaming measurements of the motion of spiral galaxies] is very marked and will require the most searching investigations in the future."

Like Slipher, Shapley made excellent measurements, but he did not make enough of them. Without a thorough understanding of Cepheids, he nonetheless tried to use them to determine the distances to objects other than globular clusters in the Milky Way. The combined work of dozens of astronomers over the next three decades, but especially the work by Walter Baade during the 1940s (Chapter Twenty), would eventually reveal that the Cepheids Shapley studied in globular clusters are intrinsically four times fainter than the Cepheids found by Leavitt in the Small Magellanic Cloud. Shapley's error was almost unavoidable. But science is self-correcting, and progress in astronomy would reveal the error made by Shapley in assuming that all Cepheids are alike.

Incredibly, despite the combination of mistakes made by Slipher and Shapley in trying to bring closure to the island universe debate, their work provided the foundation for the momentous next step in Hubble's research: the discovery that the universe was not only enormously larger than previous thought but that it was actually expanding. Conditions were ripe for an entirely new understanding of the universe and a new technique for measuring its age.

···CHAPTER 18···
The Trouble with Gravity

If there really are suns throughout the whole of infinite space, and if they are placed at equal distances from one another, or grouped into systems like that of the Milky Way, their number must be infinite and the whole vault of heaven will appear as bright as the Sun . . . .

—Heinrich Olbers, "On the Transparency of Space," 1823, as quoted in Edward Harrison, *Darkness at Night: A Riddle of the Universe* (1987)

Hubble's universe of 1925 raised an astrophysical question of enormous interest: What happens to galaxies over time in a universe made of many galaxies, massive galaxies located hundreds of kiloparsecs apart? Newton's law of gravity, formulated in 1687, had been improved upon by Albert Einstein in his 1915 general theory of relativity, but we can still use Newton's simpler conception of gravity to understand the fundamental gravity issue.

Newton's law of gravity states that the magnitude of the force of attraction between any two objects in the universe depends on the masses of the two objects and the distance between them. More massive objects attract each other more strongly than less massive objects. Objects located close to one another have a stronger mutual attraction than do objects separated by greater distances. When it comes to the galaxies in Hubble's new universe, two properties of Newton's law are most important: first, the magnitude of the force of gravity between any two objects with mass (e.g., galaxies) is never zero, no matter how far apart they are; second, that force is always attractive, pulling the two masses toward each other.

So now let's ask again, What happens to galaxies over time in a universe made of enormously massive galaxies located hundreds, thousand, or even millions of parsecs apart? For two typical galaxies separated by a few hundred kiloparsecs, the magnitude of this attractive force is enormous. Gravity inexorably pulls the galaxies toward each other. If the universe was somehow spontaneously created with galaxies separated on average by several hundred kiloparsecs and all of them sitting perfectly still in space like chocolate chips in an enormous cookie—such a universe would be called a *static universe*—then after even the smallest instant of time, each and every galaxy would feel a small gravitational tug from every other galaxy in the universe. For each galaxy, those myriad gravitational tugs, each in a different direction, would add up to a net tug in a single direction, and that direction would be toward the gravitational center, or the center of mass, of the entire universe. Each and every galaxy would begin to move, slowly at first; but then due to the relentless tugs of all the galaxies, the speeds of each and every galaxy would gradually increase. Eventually, they would all be hurtling toward the gravitational center at enormous speeds, each galaxy getting steadily closer to every other galaxy as they all rushed toward the center.

Galaxies, after all, have no choice. As physics was understood through almost the entire twentieth century, four forces existed. (In Chapter Twenty-one, we'll discuss dark energy, discovered in 1997, which acts like a fifth force in accelerating the rate of expansion of the expanding universe.) Two of these forces, the strong nuclear force and the weak nuclear force, are only important at the subatomic scale. Across larger atomic and molecular distances, electromagnetism is the dominant force, and across even larger distances, gravity rules. In a universe where gravity is the dominant force, the distances between objects—even objects scattered across the vast distances of interstellar and intergalactic space—must decrease with time. If distances are decreasing, objects are moving toward each other. Is there any alternative to this scenario? Yes, but only if the universe is infinitely big (not *really* big, but *infinitely* big), with no center and no center of mass. In an infinitely large universe every galaxy would experience equal tugs in all directions. As a result, in an infinitely large universe the net force on each and every galaxy would be zero; no galaxies would or could move at all.

Remember that in the early twentieth century, prior to January 1, 1925, the Milky Way was thought to encompass the entire universe and spiral nebulae were thought to be small objects in the Milky Way. The dominant objects in what was thought to be the entire universe were stars. If they were all moving toward each other, light from any one star would appear blueshifted when measured by an observer at the location of any other star. Therefore, an astronomer anywhere in a universe made of stars rushing toward the universe's center of mass should observe the light from every other star as blueshifted. But this was not the case. Some stars had blueshifted velocities while others were redshifted. Overall, they showed no net motion toward or away from the Earth and Sun. No evidence suggested that the objects that appeared to be the dominant constituents of the universe were rushing toward a center. If the stars were not rushing toward each other, then the universe must be infinitely large. Yet, astronomers in the early twentieth century were certain that the universe was not infinitely large. Instead, something else, some other force must exist that held the stars apart.

Olbers' Paradox

The evidence that the universe was not infinite in size derived from a centuries-old paradox., It is usually attributed to the early nineteenth-century German astronomer Heinrich Olbers, though historians have traced it back as far as 1610, to Kepler. We begin with the question, Why is the night sky dark?

Imagine a universe that is infinitely large, that has stars (or galaxies) distributed uniformly throughout its infinite space, and that is infinitely old as well. Now imagine a large but very thin spherical shell that surrounds the Earth. That shell is located at some distance, any distance, from Earth and has some arbitrary number of stars within that shell. Those stars in that single shell produce a certain amount of light. The light from all those stars travels toward Earth (as well as in all other directions). According to the inverse square law, the amount of light we receive decreases in proportion to the square of the distance from Earth to that shell. Together, the number of stars in that shell and the square of the distance to that shell allow us to calculate how much light we receive from the shell. Now imagine another concentric shell that has a diameter ten times larger, but has the same finite thickness as the first shell. With a diameter ten times larger, this second shell has 100 times more surface area and so contains a 100 times more stars than the smaller shell (since, as you will recall, stars are distributed uniformly in our imagined universe). The diameter of the second shell being ten times larger than the diameter of the first shell, each star in the outer shell is ten times more distant from Earth and thus appears 100 times fainter, as seen from Earth, than any one star in the inner shell. Therefore, the outer shell is 100 times brighter than the inner shell because it has 100 times more stars but appears 100 times fainter because those stars are ten times more distant. The net result is that the brightness of the distant shell is identical to the brightness of the nearby shell, if we assume that the universe is old enough for the light from both shells to have reached the Earth. If the universe is infinitely large, an infinite number of these shells must exist. If the universe is infinitely old, then enough time has passed for light from all the shells to have reached us. The light contributed by each and every shell, added together, becomes the brightness of our nighttime sky, and that brightness should be enormous. But the nighttime sky is dark. Something is obviously wrong with this picture of the universe.

Many possible solutions have been put forward to explain the paradox. The universe might be too young for light from all the shells to have reached us. But how does one posit a created universe that is young but infinitely large? A young, infinitely large universe was not considered plausible in 1925 (and remains physically implausible today, though a young large universe is plausible).

Or the stars and galaxies might not be uniformly distributed. Some objects might be perfectly hidden behind other objects so that light from the hidden objects cannot reach us. Even in 1925, this scenario seemed implausible in an infinite universe.

What if the universe is filled with some absorbing substance that blocks the light from distant stars and galaxies. Interstellar dust would be discovered by Robert Trumpler in 1930, but if enough dust existed to block the light from all the distant stars in an infinite universe, the dust would heat up and glow. We don't see any evidence for brightly glowing dust. Moreover, the amount of dust needed is so immense that the dust in our solar system would also obscure our own Sun. Dust cannot be the answer.

Perhaps the number of stars and galaxies in the universe is so small and they are spread so far apart that the collective light from all the stars is minimal. This argument works only if the universe is young, since even with a limited number of stars and galaxies, in an infinite universe there are an infinite number of light-producing shells. In addition, in the fairly nearby parts of the universe, we see a large number of stars. This answer, also, appears to be inadequate.

We now know that the nighttime sky is dark because the universe is both young (at least compared to being infinitely old) and finite in size, and also because the visible light emitted from distant objects is redshifted and thus does not reach us as visible light. But this understanding of the physical universe was not in-hand in 1925, let alone in 1915 when Einstein put forward his general theory of relativity. Einstein understood that in a gravity-dominated, finite-sized universe, the stars must be rushing toward each other. He was also aware that the evidence showed this not to be the case. The only sensible explanation seemed to him to be that something equivalent to a force, one that exactly opposed the attractive force of gravity and held the stars apart, must exist. He injected this anti-gravity force into his general relativity (gravity) equations through a term he called the *cosmological constant*.

The Effects of a Finite Universe

The year is now 1926. Shapley's universe has given way to Hubble's universe. Most of the stars we see are part of and presumably orbit the center of the Milky Way Galaxy. Galaxies, not stars, are now known to be the most massive and therefore the gravitationally dominant objects in the universe. The evidence Einstein considered a decade earlier—

that the stars, overall, showed no net motion toward or away from the Earth and Sun—is no longer relevant. Now we need to wonder about the velocities of galaxies. Again, we ask, What happens to galaxies over time in a universe made of massive galaxies located hundreds of kiloparsecs apart? The obvious answer: if they are moving closer together, then they should all show blueshifts in their spectra.

If the universe is not infinitely big then we have to consider other possibilities. Perhaps the universe is very young, so young that the galaxies have not yet been accelerated to appreciable velocities. Or perhaps there is an unknown force that acts in such a way as to exactly oppose the long-distance attractive nature of gravity. Can we now posit that the cosmological constant applies to galaxies rather than stars?

When it comes to figuring out if enough time has passed for the force of gravity to have asserted its will on the motions of galaxies, the exact age of the universe does not matter, provided it is large. For astronomers, who by the 1920s knew that the Earth and Sun, and therefore the universe, were at least a few billion years old, this was a testable hypothesis.

How does gravity impose its will on objects? Forces make objects move. An object that is at rest but subject to the force of gravity will start moving. The change from not-moving to moving means that the motion or velocity of the object has changed; in the language of physics, such a change, or more specifically the rate at which the velocity of an object changes over time because a force is acting on it, is known as *acceleration*. Forces accelerate objects, but a force of a given strength does not accelerate all things equally. A given force will impose a large acceleration on a small mass but only a small acceleration on a big mass.

Let's examine how the force of gravity due to the Earth attracting a person accelerates that person downwards. If the solid ground did not get in the way (for example, if the person jumped out of a hot air balloon without a parachute), the Earth would accelerate this suicidal skydiver downwards at a rate of about 10 meters per second per second. This means that the velocity at which a person would be pulled downwards increases by 10 meters per second with every passing second. After one second, the person's downward speed would be 10 meters per second; after two seconds, the downward speed would have increased to 20 meters per second; after three seconds, the downward speed would have increased to 30 meters per second. If we think about the force of gravity in these terms, we can compare the acceleration induced by the Milky Way on a distant galaxy to the acceleration induced by the Earth on a person standing on the surface of the Earth. When we do this, we find that even though the galaxy-on-galaxy forces are enormous, so are their masses. With such large masses, they are very hard to accelerate. Consequently, the acceleration induced by one galaxy on another is thousands of billions of times weaker than the acceleration induced by the Earth on a person who has jumped into the air from the surface of the Earth.

Nevertheless, although the accelerations in the relative motions of galaxies are very small compared to the effect of the nearby Earth on an object near the surface of the Earth, given enough time, real, measurable motions will ensue. At first the velocities will be immeasurably small. But with every passing second, the forces will continue to act and the galaxies will continue to accelerate. After a million years, these velocities could reach tens of meters per second; after a few billion years, the relative velocities of the galaxies would be several tens of kilometers per second. Since, a century ago, astronomers knew that the universe was at least several billion years old, they would have expected the galaxies to be flying toward each other at typical speeds of several tens of kilometers per second. Were they?

The Motions of Galaxies

From 1912 to 1914, Vesto Slipher had made observations that, though not intended for this purpose, provided the answer to that question. The great spiral nebula Andromeda and its neighbor M32 were moving toward the Sun at speeds of 300 kilometers per second; but the nine other spiral nebulae whose velocities Slipher could measure had speeds ranging from 200 to 1100 kilometers per second *away* from the Sun (these speeds would be comparable if measured toward or away from the center of the Milky Way rather than toward or away from the

Sun). Excluding Andromeda and M32, the average velocity was 550 kilometers per second away from, not toward, the Sun. By 1917, Slipher had expanded his research sample to include about thirty spirals, for which he obtained an average velocity of 570 kilometers per second away from the Sun.

Remember that in 1917 the Milky Way was thought to encompass the entire universe and spiral nebulae were thought to be objects in the Milky Way. Spirals may have had, collectively, unusual velocities (large and directed away from the Sun) compared to stars, but probably few astronomers and even fewer physicists were aware of Slipher's measurements. The dominant objects in what was thought to be the universe were stars, and stars had much smaller velocities (tens of kilometers per second), some positive, some negative, with no net motion toward or away from the Earth and Sun. Few in 1917 considered the possibility that the spirals were galaxies, let alone that these pieces of the universe were flying apart.

After New Year's Day, 1925, Slipher's measurements of the velocities of spiral nebulae took on a new importance. Almost overnight, galaxies, not stars, became the dominant individual constituents of the universe. Applying Einstein's logic to the spirals, we should expect them to be hurtling toward each other at measureable speeds, but they were doing just the opposite.

Mathematically, Einstein's equations for gravity, for general relativity, permit many solutions. His original understanding of his own equations was such that, when solved, they permitted either a universe that contracted under the force of gravity or a universe that, with the introduction of the cosmological constant, remained static. But, as it turns out, Einstein's equations permit other solutions. A young Russian mathematician, Alexander Friedmann, discovered in 1922 that they can also describe an expanding universe. Einstein's equations could, in other words, be consistent with a universe in which most of the galaxies were hurtling away from each other. But before Friedman's discovery could be applied to the problem of an expanding universe, astronomers had to become aware that the problem existed.

CHAPTER 19

The Expanding Universe

If I have seen farther, it is by standing upon the shoulders of giants.

—Isaac Newton, in a letter to Robert Hooke, February 5, 1675, as quoted in David Brewster's *Memoirs of the Life, Writings, and Discoveries of Sir Isaac Newton* (1855)

Until very recently in human history, those who thought hard about the universe, whether they were theologians, natural philosophers, or scientists, could agree on one thing. The universe had a center. In the Aristotelian universe, which was viewed as consistent with and supportive of medieval biblical theology, the Earth occupied the privileged center of a small and finite universe, with the sphere of the stars just beyond the sphere of Saturn and, in the view of many cognoscenti, with the spheres of the angels and of God in the seventh heaven, just beyond the celestial sphere. In 1543, Nicholas Copernicus set off a century-long revolution that relegated Aristotle's celestial spheres to the dustbin of history: in the Copernican universe, the Sun assumed the central throne in a universe that was understood to be bigger than the Aristotelian universe, but only slightly bigger. It had to be just big enough to accommodate the stars at distances too great for their parallax angles to be measured.

In 1917, Harlow Shapley shattered the seemingly self-evident notion that our nearby, all-important Sun was at or near the center of the universe; in Shapley's universe, the center of the Milky Way, a location that he measured to be fully 13 kiloparsecs (40,000 light-years) away from the Earth and Sun and utterly irrelevant to life on Earth, emerged as the new center. In Shapley's view, the Milky Way

was the entire universe and its size, though large, remained compre-
hensible: 90 kiloparsecs (300,000 light-years) from end to end. In
1925, Edwin Hubble's discoveries of Cepheids in both Andromeda
and Barnard's Galaxy stretched the size of the measured universe to
millions of parsecs, but for all its extraordinary size it apparently still
had a center. Soon though (in the 1920s), Hubble would shatter even
that centuries-old "truth."

Receding Galaxies

By 1929, Hubble, working with his assistant Milton Humason, had
begun a project designed to measure distances to extragalactic nebulae.
He started with a target list of forty-six galaxies for which he had radial
velocities in hand. Forty-three of these were galaxies for which Slipher
had already obtained radial velocity measurements. (Slipher's 1917
survey of thirty spirals had expanded to forty-three by the time Eddington compiled the list that he published in his great 1923 treatise *The
Mathematical Theory of Relativity*.) By 1929, Humason had obtained
spectra and measured the spectral velocities for three additional extragalactic nebulae not previously measured by Slipher.

Hubble was able to identify Cepheids in six of these forty-six galaxies. For each galaxy with a Cepheid, Hubble calculated a distance using
Leavitt's period-luminosity relationship, as calibrated by Shapley. With
that distance in hand, he calibrated the average absolute brightness of
the brightest individual objects that looked like stars (which he thought
were stars but acknowledged could have been remote star clusters) in
that galaxy. The result? The average absolute brightness for the brightest star-like objects in each of those six galaxies was the same.

Hubble next made an assumption about galaxies analogous to the
assumption that Shapley had made for globular clusters a decade earlier. He reasoned that, since the average absolute brightnesses of the
brightest star-like objects in the six galaxies for which he had measured
the absolute brightnesses were identical, the average absolute brightness of the brightest of these objects in every spiral galaxy should be
identical. If the brightest star-like objects in every galaxy are identi-

cally bright they become, in the language of modern astronomy, *standard candles*, and they can be used to calibrate distances. How?

First, you measure the average apparent brightness of the brightest star-like objects in a galaxy. Then, use the absolute brightness for these objects obtained from the standard candle measurements of the first six galaxies together with the inverse square law for light to calculate the distance to the galaxy. Hubble was able to identify the brightest star-like objects in eighteen of the forty other galaxies in his target list. After measuring the apparent brightnesses of these, he was able to calculate distances to these galaxies, giving him twenty-four galaxies with measured distances.

In January of 1929, Hubble published a paper on these twenty-four nebulae entitled "A Relation Between Distance and Radial Velocity among Extra-Galactic Nebulae." In it he proposed "a roughly linear relation between velocities and distances among nebulae," a proposition that we now call Hubble's law and write in the form $v = H_0 d$. Hubble's law tells us that a linear relationship exists between the redshift velocity (v) of the galaxy and the distance (d) to the galaxy. The redshift velocity and distance are related, mathematically, through the constant of proportionality that we call the Hubble constant (H_0). Once the value of H_0 has been measured from Hubble's observations of v and d for the first twenty-four galaxies, the distance to any galaxy can be calculated by measuring the galaxy's redshift velocity from an observation of the spectrum of that galaxy and then solving Hubble's law for the unknown distance.

If one galaxy has a redshifted velocity that is twice as large as that of a second galaxy, Hubble's law informs us that the distance to the first is twice the distance to the second. The value of the constant of proportionality, H_0, tells us the actual physical distances to both galaxies, provided that we have made measurements of their redshifted velocities. The velocities of the galaxies in Hubble's first survey ranged up to 1,090 kilometers per second. According to Hubble's initial calibration of H_0, these velocities correspond to distances of up to about two million parsecs (two megaparsecs, or 6 million light-years). In 1929, Hubble concluded that it is "premature to discuss in detail the obvious consequences of the present results."

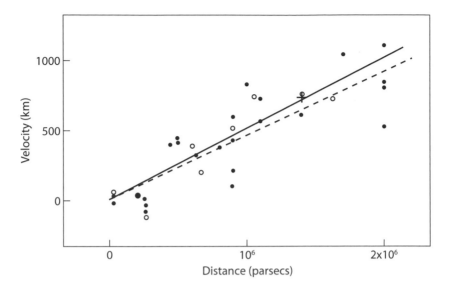

Figure 19.1. Hubble's 1929 plot showing that the distance to a galaxy is directly proportional to the redshifted velocity of that galaxy. Distance is measured along the x-axis, with each box representing one million parsecs. Velocity is measured along the y-axis (each box representing a speed of 500 km per second. Note that the original labels on the y-axis have incorrect units and, as we will see in Chapter Twenty, the x-axis has the wrong values). From Hubble, *Proceedings of the National Academy of Science* (1929): 15, 169.

In 1931, Hubble and Humason dramatically extended their reach out into the universe. They reported velocity measurements as large as 19,700 kilometers per second for redshifted galaxies, putting the most remote galaxy in their survey at a distance of thirty-five megaparsecs (110 million light-years), about 18 times farther out into the universe than had been possible only two years earlier. The linear relationship between redshift velocity and distance still held, with the value of H_0 placed at 560 kilometers per second per megaparsec. With this value of the Hubble constant, astronomers now had a tool with which they could calculate the distance to any galaxy in the universe, provided they could measure the redshift velocity of that galaxy. At a distance of one megaparsec, a galaxy should have a redshift velocity of about 560 kilometers per second; at a distance of two megaparsecs, a galaxy should have a redshift velocity of about 1,120 kilometers per second; and at a

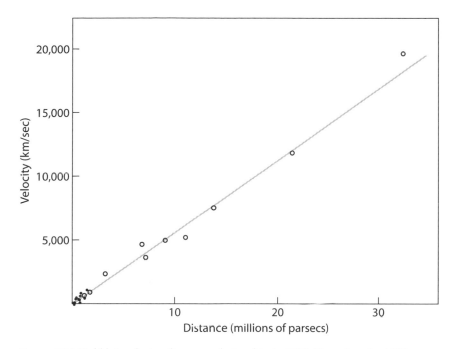

Figure 19.2 Hubble's velocity-distance relationship in 1931. Note that the 1929 measurements extended out to a distance of only two million parsecs, i.e., the leftmost fifth of the leftmost and lowest box. Each open circle represents the mean value for a cluster of galaxies. From Hubble and Humason, *The Astrophysical Journal,* (1931): 74, 43.

distance of ten megaparsecs, a galaxy should have a redshift velocity of about 5,600 kilometers per second. Therefore, if we measure a galaxy as having a redshift velocity of 28,000 kilometers per second, it must lie at a distance of 50 megaparsecs.

Interpreting Hubble's Law

Hubble and Humason were very careful in their 1929 and 1931 papers to present their results empirically, "without venturing on the interpretation and cosmological significance." They even suggested that "the interpretation of redshifts as actual velocities . . . does not inspire the same confidence [as the association of apparent and absolute magnitudes with distance], and the term 'velocity' will be used for the present in the sense of 'apparent' velocity, without prejudice as to its ultimate significance." Hubble left the interpretation of his measurements to others, to us.

How are we to interpret Hubble's law? We look outwards from the Earth (i.e., from the Milky Way) and we see that all other galaxies (other than our very closest neighbors and Andromeda and its companion galaxies) appear to be moving away from us; we measure the speeds at which these galaxies appear to be rushing away from us and notice that these speeds increase as the galaxies' distances from us increase. What does this correlation between distance and velocity away from us mean? One explanation appears obvious to most of us:

1. The universe exploded like a bomb and all the fragments flew away from the center *through space*. Space itself, which existed before the explosion but had been empty except for the tiny unexploded bomb at its center, is large enough to encompass for all time the outward-flying pieces of the explosion. Not all pieces of debris were ejected from the explosion site with the same velocity so, in any fixed time interval since the explosion, the fastest-flying pieces have flown the farthest and the slowest-flying pieces have flown the shortest distances. All other pieces have traveled intermediate distances in proportion to their speeds. After any amount of time, a stationary observer *looking outwards from the center* would measure a positive, linear correlation of distance with velocity for the expanding debris cloud, just as Hubble found for the galaxies. In this universe, space is unchanged while the *galaxies move through space*. In this universe, the redshifts of galaxies are actual Doppler shifts, caused by the physical motion of the galaxies away from us through space.

A second explanation is much harder to grasp but fits the observations equally well:

2. The galaxies are fixed at specific locations in space and do not move through space. Over time, the very fabric of *space is stretched*

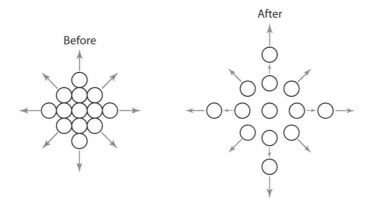

Figure 19.3. Illustration of exploding bomb fragments. Left, all fragments begin at (or near) the center. Right, after the explosion, the fastest moving fragments have moved the greatest distance from the center, while the objects at the center are not moving at all.

uniformly so that the separations between galaxies increase with time. The galaxies themselves are not stretched. (In this view of the universe, on a local scale— "local" meaning distances within galaxies rather than distances between galaxies—electromagnetic and gravitational forces are stronger than the force that is stretching the universe. Consequently, the galaxies, the stars, the solar system, the Earth, and your pet cat are not being stretched by the stretching of space.) The total increase in separation between galaxies increases in direct proportion to the initial separation between them. Thus, a stationary observer *located anywhere* in this universe would measure a positive, linear correlation of distance with apparent velocity for the apparent recession of distant galaxies, which is exactly what Hubble found. In this universe, *space is stretched but galaxies do not move through space.* In this universe, the redshifts of galaxies are not Doppler shifts since the galaxies themselves are not moving through space. Instead, we call these redshifts cosmological because the photons emitted by the non-moving galaxies are stretched—redshifted—as they travel through the continuously expanding universe.

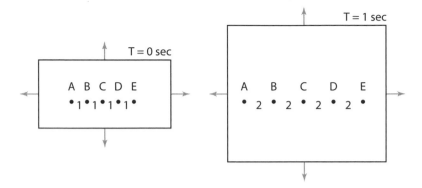

Figure 19.4. Illustration of Hubble's law. Left, at time 0 seconds, we have a piece of paper with five dots (labeled A, B, C, D, and E). Each dot is separated from the next closest dot by one unit of distance (e.g., one meter). Right, one second later, our piece of paper has been stretched in all directions so that it has doubled in size. The distance from A to B has doubled from one to two meters. The distance from A to E has doubled from four to eight meters. An observer at dot A would discover that dot B is moving away from dot A at a speed of one meter per second, while dot E is moving away at a speed of four meters per second. An observer at any position would find that the other dots are moving away from that position with speeds that are directly proportional to their distances from that position.

The Age of the Universe Reckoned from the Hubble Expansion

At least one conclusion that is common to both of these interpretations of Hubble's law amounts to a cosmological revolution: the universe had a beginning in time that can be measured by scientific means. How so?

No matter which of the two interpretations of Hubble's distance-redshift relationship you prefer, if we look backwards in time, the galaxies are closer together, either because they are moving backwards through space toward the center or because space itself is getting smaller. If we look back far enough, all the galaxies merge together into a single, super, pregalactic clump. If we continue backwards in time beyond the moment when this pregalactic clump formed, all the mass in the pregalactic clump merges into a single point. Consequently, to

infer the age of the universe we need only imagine reversing the motions of the outrushing galaxies and then calculating how much time is
needed before they all smash together at the beginning of time.

Imagine you are driving a car eastward from your home city on an
interstate highway. You set the cruise control at a speed of 100 kilometers per hour. I am an observer in your home city, using a radar detector
to track your car. I do not know when you started your trip, but at any
moment I can measure your velocity (speed in a specific direction) and
your distance from home. Since I know that distance equals velocity
multiplied by time, I also know that the time since you started your trip
is the distance divided by the velocity. Thus, I can use my measurements of your velocity and your distance from home to calculate the
length of time required for your trip, so far. If you are traveling at a velocity of 100 kilometers per hour and have traveled 400 kilometers,
then you left home four hours ago. This calculation is strictly true provided the speed of your car is unchanged from the moment the trip
began until the moment I used my radar system to measure the speed of
your car.

Now imagine that you and a second driver leave your home city on
road trips at the same moment but are driving your cars at different
speeds. If the second driver is driving at 50 kilometers per hour and has
traveled 200 kilometers, I find that this second driver has also been
traveling for four hours. You and the second driver left on this trip at
the same time, just like runners in a marathon who all start at the same
moment but run at different speeds and so end up at different distances
from the starting point at any given instant. If I were to graph your
velocities (x-axis values) versus your distances traveled (y-axis values),
the first data point (50 km per hour; 200 km) and second data point
(100 km per hour; 400 km) would fall along a straight line on the plot
that extended through the origin (0 km per hour; 0 km). The slope of
this line (the y value divided by the x value), four hours, is the constant of proportionality, a sort of Hubble constant, for this set of drivers. It is also the time since both drivers began their trips.

Notice in this analogy that if we had waited to make all of our measurements until after the first driver had driven 800 kilometers and

the second driver had driven 400 kilometers, we would have obtained a straight line on our plot, but the data points would have been at (50 km per hour; 400 km) and (100 km per hour; 800 km). Thus, the slope of our line, our Hubble constant, would have a different slope: eight hours. In other words, the Hubble "constant" changes with time. It is not a constant; it is a time that increases steadily the longer the drivers drive.

If Hubble's data for the distances and velocities of galaxies were plotted with velocity on the x-axis instead of the y-axis and distance on the y-axis instead of the x-axis, the slope of the line drawn through the data points would be the Hubble constant, and the value of the Hubble constant would be the current age of the universe (t_H, which is known as the *Hubble time*), assuming that the rate at which galaxies are moving through space or the rate at which space has been expanding has been uniform, unchanged since the beginning of time. Even as early as 1931, however, cosmologists had calculated that the expansion rate of the universe had not been constant over its entire history. The age of the universe therefore was not identical with the Hubble time; instead, from Einstein's theory of general relativity, they derived an actual age for the universe that was two-thirds of the Hubble time.

Astronomers have a habit of doing things backwards, inside out, or upside down (often, but not always, with good reason). When Hubble plotted his data for the distances and velocities of galaxies, he put the distances on the x-axis and the velocities on the y-axis. As a result, the slope of the plotted line is the inverse of the Hubble time rather than the Hubble time itself. If we had plotted the data for our cars in this way, the slope of the line (y divided by x) would be 0.25 kilometers per hour per kilometer, or simply 0.25 per hour. The inverse of 0.25 per hour is 4.00 hours, so we can calculate the Hubble time from the Hubble constant by flipping the Hubble constant upside down ($t_H = 1/H_0$). And as we discovered from our cars and drivers, since the Hubble time must increase with time, the Hubble constant would decrease with time in a universe in which the apparent velocities of galaxies never change.

Are Galaxies Moving through Space or Is Space Expanding?

If the only pieces of information we had about the universe were the distances and recession velocities of galaxies, could we distinguish between a universe in which the galaxies were moving through space and a universe in which space was expanding? In both models, an observer placed almost anywhere in the universe and riding with one of the galaxies would make the identical observations that virtually all objects move away from the observer and that nearby objects recede more slowly than distant objects. The physical principles that underlie these observations are very different, however. To distinguish between these alternative explanations, we need to gather more information.

Imagine that the universe is shaped like an immense loaf of raisin bread, with each raisin representing a galaxy. And each raisin has a tiny mathematically astute observer riding with it who can make the necessary distance and redshift velocity measurements for our experiment. Perhaps the loaf is expanding as it bakes, so the raisins are carried outwards by the expanding dough, or perhaps the raisins, via some unknown mechanism, are moving outwards through the dough toward the crust. Some of the raisins are located so far from the surface that no matter how diligently their observers observe, they would never be able to see as far as the edge; they could never demonstrate that they were inside a loaf of bread that has a crust, with the crust being the boundary between bread and not-bread. For all observers located on raisins embedded deeply inside the loaf of bread, the universe would look the same in all directions and for as far as they could see. These observers would eventually discern the existence of the Hubble equation and discover that it applies in all possible directions in which they might choose to look.

Some raisins, however, are close enough to the outside of the loaf that they can see to the crust and beyond. These near-the-crust observers, after making distance, velocity, and position measurements of other raisins, would discover a number of differences between what they see and what the far-from-the-crust raisins observe. First, they would note that they could see far more raisins in one direction than in the other. Second, they would discover that their universe had an abundance of raisins located at great distances from their location as seen in one direction, but a dearth or complete absence of distant raisins in many other directions. While the Hubble equation would apply to all raisins observed in all directions, in the direction toward the nearby crust the Hubble equation would only apply out to a limited distance, beyond which there would be no raisins and therefore no objects against which the observer could test Hubble's law.

The Cosmological Principle

What about us? What about the observations that astronomers riding on Earth make from inside the Milky Way Galaxy? For us, the Hubble equation appears to apply to galaxies for as far as we can see in all directions. We could assume that the Milky Way just happens to be far from the edge of the universe, like one of the raisins located deep inside the loaf of raisin bread; under this assumption, we, the residents of the Milky Way, occupy a central or nearly central point in the universe (or, as it were, in the loaf of raisin bread).

Since the time of Copernicus, however, astronomers, physicists, and philosophers have come to reject the idea that we live at a privileged place in the universe. This principle, known as the *cosmological principle*, has grown in acceptance over the centuries until it is now one of the most important pillars of modern thought. According to the cosmological principle, *the laws of physics must be the same everywhere in the universe.* This assumption, that the laws of physics are universal, is called *universality*. A second assumption upon which the cosmological principle is built, which is implicit if no place in the universe is privileged and if the laws of physics are universal, is that on large spatial scales *the universe looks the same in all directions to all observers everywhere in the universe.* This second assumption is known as *isotropy*. Finally, if the cosmological principle is valid, and if we accept the assumptions of universality and isotropy, then averaged over large spatial scales *the universe must have the same average contents everywhere.* This third assumption, known as *homogeneity*, implies that any moderate-sized piece of the universe contain the same kinds of things in it as any other, whether we mean large-scale entities (stars, galaxies, galaxy clusters) or microscopic ones (the relative amounts of the various elements, the numbers of protons and electrons). If universality, isotropy, and homogeneity are all valid concepts, then no parts of the universe should be in any way special or privileged. The traditional loaf-of-raisin-bread model, having both a center and a crust, violates the cosmological principle by violating the assumption of isotropy: the universe does not look the same in all directions from all locations in the raisin bread model.

The only way in which all observers located anywhere in the universe (or in that confounded loaf of raisin bread) would all observe the same thing—that is, the only way that the Hubble equation would apply to all galaxies (raisins) seen out to all distances in all directions—would be if the universe (loaf of bread) had no edge (crust). For now, we will assume that the cosmological principle is valid in the universe; that is, that the assumptions of universality, isotropy, homogeneity are all correct. Later (in Chapter Twenty-five), we will discuss the experimental and observational evidence that supports the cosmological principle. For now, let us simply accept the cosmological principle and we will find ourselves well along our way to understanding how to interpret Hubble's equation. Our next step: we must figure out how to construct a universe without an edge.

A Universe without an Edge

Conceptually, one way to make a universe without an edge would be to make a universe that is infinite in size. But remember that the resolution of Olbers' paradox strongly argues that the part of the universe containing luminous matter that we can see is finite. This, combined with the cosmological principle, forcefully suggests that we do not live in a universe that is infinite in both time and size. (It does not, however, rule out the existence of other dimensions or of some form of multiverse).

The second method of solving the problem is for our loaf of bread to have a shape that closes on itself. For example, imagine a loaf of bread that is shaped like the leather surface of a basketball. Conceptually, this loaf of bread requires a great deal of effort to understand (not to mention the difficulties that would be involved in baking it) because when we think of a basketball, we tend to think about the three-dimensional shape of the ball and its environment: it has an inside filled with air (not leather) and is surrounded on the outside by air (not leather). The leather surface is fairly thin compared to the size of the ball. For the purposes of our analogy, this way of thinking about the basketball is flawed. We must instead think of the curved leather surface as three-dimensional space itself; there is no inside or outside that can be located *in three-dimensional space*. If you are an ant walking on the surface of the basketball, you cannot walk into the interior or jump off into space; any attempt you make to explore space will be done by walking on and staying on the surface of the basketball. Similarly, in our curved-space universe, any measurements we make of space (velocities and distances of galaxies) must be made along the curved surface of the universe, because those are the only directions in which our three-dimensional space exists; the inside and outside "directions" do not exist in three-dimensional space (if they did, then our three-dimensional universe would have both an inside edge and an outside edge, which we have already determined is impossible). The inside and outside of our basketball do, however, exist *in time*. We will develop this basketball analogy in more depth later on.

Think of the universe as an infinitely big volume of empty space. Then, at a single moment, which we will call the beginning of time, all the galaxies in the observable universe appear in a single, tiny region of that universe. If the motions and distances of galaxies are well described by the Hubble equation, then at the next moment the galaxies will have begun to move apart. If the most distant outrushing galaxies moved *through space* at speeds less than infinitely fast, then an outer distance limit would exist for this expanding cloud of galaxies, beyond which no galaxies could exist. But we have already concluded that the universe has no edge, no outer limit beyond which no galaxies exist; otherwise, there would exist a moment in the history of the universe—

which could be the present moment or any other time—when some observers would and others would not see the boundary beyond which no galaxies could be found. Therefore, in order to preserve the cosmological principle, the universe must have instantaneously become infinitely large. Such an event would require that an infinite number of galaxies moved *through space* at infinite speeds for a brief period of time in order to reach their current locations. This explanation contradicts the laws of physics, since no objects can move *through space* faster than the speed of light, let alone infinitely fast; this explanation cannot be correct.

Now ignore the question of how the universe started and became infinitely big and just think about a universe that is infinitely big and is well described by the Hubble equation. In this universe, distant galaxies travel away from us at greater speeds than do nearby galaxies. The distance between us and some galaxies is increasing at the speed of 10,000 kilometers per second; beyond these galaxies are others that are receding from us at speeds of 50,000, even 100,000, or even 200,000 kilometers per second. Some galaxies must lie at distances so great that their recession speeds—remember, we must think of these speeds as the motion of the galaxies *through space*—would exceed the speed of light, but this cannot happen. Based on decades of experiments, physicists in the early twentieth century reached consensus that the speed of light is finite. In a vacuum it is clocked at 299,792.458 kilometers per second. Albert Einstein, in his 1905 theory of special relativity, enshrined two related principles in our understanding of the physical universe: the speed of light is absolutely the same for all observers in the universe; and no objects in the universe can move *through space* faster than the speed of light. (Einstein's axiom for the speed of light places no such constraint on the expansion of space itself, only on the movement of objects through space.) More than a century has passed since Einstein enunciated these principles, and thousands of experiments done in laboratories all over the world have repeatedly and convincingly affirmed and reaffirmed his ideas about the speed of light. Yet, in an infinite universe filled with galaxies in which Hubble's law correctly describes the motions of galaxies and in which the cosmological principle holds true, almost all galaxies must lie at distances so vast that their recession speeds *through space* would exceed Einstein's speed limit. An infinite universe could not be governed by Hubble's law *and* by the theory of special relativity *and* by the cosmological principle. As strange as the ideas in relativity were a century ago, they had by 1929 become foundational pillars for modern physics and today, after almost a hundred years of continuing experiments and theoretical developments, those pillars are as sturdy as ever. With a choice between either Einstein's physics or an infinite universe in which privileged observers live and objects move *through space* faster than the speed of light, Einstein's physics wins.

A Finite Expanding Universe

We know now that our universe has no edges, that the part of it that includes luminous, observable galaxies is finite, and that the galaxies obey Hubble's equation. But how do we put these ideas together and paint a coherent picture of such a universe?

Shortly after Einstein published his general theory of relativity in 1915, the Dutch astronomer Willem de Sitter attempted to solve Einstein's equations. De Sitter's mathematical solution was unphysical because he included no ordinary matter or light and so no source of gravitation across enormous cosmic scales in his equations, but he did discover through his calculations a different way to interpret the spectral lines of distant galaxies: the *stretching of the space* of the universe, he decided, would also stretch the photons. A blue photon when stretched becomes longer and therefore redder. If space is being stretched, then light from more distant objects will be stretched more than light from nearby objects. Hubble's law, not yet observed when de Sitter worked out his solution to Einstein's equations, is indicative of *the stretching of space, not of galaxies moving through space*. In 1931, de Sitter's concept of stretching space provided the explanation of Hubble's observations that astronomers were seeking: galaxies are not moving through space. Rather, space itself and the photons traveling through it are being stretched while the galaxies remain approximately fixed in their locations in space.

This idea bears repeating because, although many of us have seen pictures or cartoons of the expanding universe, few people other than professional astronomers and cosmologists have truly grasped what is meant in this explanation of it. So here it is again: Space that is empty of matter does not and never did exist. The distances between galaxies, galaxies being dense pockets of matter in an otherwise rarefied universe, are increasing with time but the galaxies themselves are *not* speeding *through space*. On the grandest scale (that is, ignoring small local motions that all galaxies have as they interact with nearby galaxies), galaxies are fixed in their locations in space. They do not move through space. Yet the distances between galaxies grow rapidly because the fabric of space in between the galaxies is continually being stretched, taffy-like, with the galaxies patiently riding along at fixed locations in the taffy.

In a March 1948 radio broadcast on the BBC show *The Nature of Things*, British astrophysicist Fred Hoyle used the term *Big Bang* when he described for his listeners this model of the expanding universe. Though Hoyle spent considerable effort over more than five decades challenging the Big Bang theory—he preferred the steady state theory, which envisions the universe as eternal and unchanging—Hoyle's name for the theory of the expanding universe stuck.

A Universe with No Center

Now let's return to our basketball analogy. The leather skin of the basketball is three-dimensional space, the fabric of the universe. The basketball universe began as a single point, then expanded. The universe expanding is analogous to our basketball inflating, but with one all-important difference: in our experience, a basketball exists within space and expands through space while in our analogy the basketball is space. In our analogy, the "space" outside of the basketball does not exist, at least not today and not in any of our three dimensions, because today it is not part of the universe (which is limited to the skin of the basketball). The inside of the basketball does not exist today, in three-dimensional space, for the same reason. We could not travel to the inside or outside of the basketball by traveling through three-dimensional space, even if we could travel infinitely fast. The inside of the basketball exists only in the past, as a smaller basketball, while the outside of the basketball exists only in the future, as a larger basketball. Since time travel is not (yet) possible, we are absolutely limited to traveling only through space. We are restricted, by the laws of physics, to the skin of the basketball.

Even though we cannot travel into the past, astronomers can look into the past. Light, even traveling at speeds of nearly 300,000 kilometers per second, requires measurable lengths of time to cross the vast distances of the universe. Light from an object located 4 million light-years away would need 4 million years to cross the space of the universe between it and us (if we ignore the fact that space is expanding while those photons are traveling across the universe); therefore, the light reaching us today should have left that object 4 million years ago, at which time the universe was smaller. Therefore, when astronomers look at distant objects, they are looking through space and also backwards across time. Astronomers see distant objects as they were when the observable universe was both younger and smaller; they look toward the inside of the basketball, into the past. In effect, telescopes are time machines, and astronomers are permitted to look at but not to touch the past.

···CHAPTER 20···

The Hubble Age of the Universe

Hubble took the more cautious line that one should not overwork the principle of uniformity and that there might be a real difference between the richest globular clusters of the Andromeda nebula and those of our own galaxy.

—Walter Baade, in "The Period Luminosity Relation of the Cepheids," *Publications of the Astronomical Society of the Pacific* (1956)

We now have a model with which to understand the history of the universe: the universe began with all its matter and energy very close together, because space itself was vanishingly small. Over time, space has expanded. The galaxies, which formed within a few hundred million years after the birth of the universe, have remained fixed at their original locations in space. It is on account of the expansion of space that the galaxies appear to be flying apart even though they are not moving through space. The distances between galaxies are directly proportional to the speeds at which they appear to be flying apart, so that galaxies now separated by large distances appear to be flying apart at greater speeds than do galaxies that are closer together. The relationship between the separation distances and the recession speeds of galaxies is what we call Hubble's law. The constant of proportionality in this relationship, which tells us how many megaparsecs separate galaxies for every kilometer per second of velocity by which their speeds differ, we call the Hubble constant, H_0. The Hubble constant directly gives us the Hubble time, which in turn gives us an estimate for the age of the universe.

Hubble's work of 1931 gives a Hubble time of approximately 1.8 billion years or an age of the universe, estimated at that time to be about two-thirds of the Hubble time, of about 1.2 billion years. Roughly speaking, this age is consistent with the then-estimated age of the Earth. "The age of the crust is probably between two and eight thousand million years," concluded none other than Henry Norris Russell in 1921. By 1930, however, geologists were convinced from radioactive dating of rocks that the lower limit to the age of the Earth was 3 billion years. If radioactive dating was a valid tool and Hubble's measurements were accurate, then the universe was too young. It was billions of years younger than its constituents. This paradox became known as the *time scale difficulty* or, in the words of historian of astronomy John North, "the nightmare of cosmologists." Something was wrong.

The Limitations of Ockham's Razor

In 1931, after calibrating the distance to the Andromeda Nebula using Cepheids and the period-luminosity relationship, Hubble had noticed that the brightest globular clusters in Andromeda were four times (1.5 magnitudes) fainter than the brightest globular clusters in the Milky Way. By 1940, this discrepancy had been confirmed by other astronomers and had become a source of discomfort. Why should globular clusters in Andromeda be so different from globular clusters in the Milky Way? If the two galaxies had very different kinds of stars in their globular clusters, then the principle of uniformity was invalid; however, if the Milky Way and Andromeda have similar stars that obey the same laws of physics, then the calculated distance to Andromeda must be incorrect.

Shapley, in his work on Cepheids, had included all Cepheid-like variable stars together into a single period-luminosity relationship, and it was Shapley's calibration that Hubble used. What neither Shapley nor Hubble knew is that not all Cepheid-like variable stars are alike.

The H-R diagram has a region we now call the *instability strip*. The instability strip extends from the main sequence upwards (toward higher luminosities) and to the right (toward cooler temperatures). Stars in this quadrant pulsate, alternately getting bigger, brighter, and cooler and then smaller, fainter, and hotter. These pulsations drive the period, brightness, and color (spectral type) changes observed in "pulsating variable stars," of which there are three important types:

- Classical Cepheids, also known as Delta Cephei stars or Type I Cepheids, which are about four times brighter than Type II Cepheids. These stars have pulsation periods ranging from two to forty days, luminosity amplitudes of about one to one-and-one-half magnitudes, and intermediate temperatures and spectral types (F, G). They are found in the spiral arms of spiral galaxies and in the Small Magellanic Cloud; they are not found in globular clusters. Examples include Polaris and Delta Cephei, itself;
- W Virginis stars, or Type II Cepheids (the ones described well by Shapley's work on globular clusters), which are about four times fainter than Type I Cepheids. These stars also have pulsation periods ranging from two to forty days. They are found in the halos of spiral galaxies, in elliptical galaxies, and, unlike Type I Cepheids, in globular clusters;
- RR Lyrae stars, or cluster-type variables (the ones discovered by Bailey), which have pulsation periods of only 0.3 to 0.9 days, luminosity amplitudes of about one magnitude, and intermediate temperatures and spectral types (A, F). They are found in globular clusters and in the halos and nuclei of spiral galaxies.

When Shapley put all three types of stars together to come up with a single period-luminosity relationship, he committed the error of mixing apples, oranges, and pears, which is a common and almost unavoidable problem in astronomy soon after the initial discovery of a new kind of object.

When only a few objects are known that share a particular trait, and when astronomers know very little about these objects, the similarities among them very naturally are considered of primary importance, while differences are only secondary. If the objects are of sufficient interest, they will excite considerable study and, as a result, a large amount of data about these objects soon becomes available. In the next phase of understanding, astronomers begin to discern differences between the objects that may be significant enough to warrant the establishment of subtypes or subclasses. Just as a shortage of data about the properties of nebulae led Slipher to assume they were all alike, the absence of a large enough body of data on Cepheids and cluster-type variables prevented Shapley and others in the 1910s and 1920s from recognizing the systematic differences between the several types of periodic variable stars they had lumped together at that time as Cepheids.

These mistakes illustrate the dangers of applying the principle of Ockham's razor in science. William of Ockham was a fourteenth-century philosopher who argued that what transcends experience cannot be verified by rational, logical arguments. Ockham's work of course predated the birth of modern science (his primary interest was in proofs of God's existence), but modern science has embraced his dictum on the value of simplicity, known as Ockham's razor: What can be done with fewer assumptions is done in vain with more. That is, do not needlessly overcomplicate a scientific explanation. Keep it simple. In the case of Cepheids, Shapley's assertion in 1918 "that variables, of a given period are of comparable luminosity, whether they are in the general galactic system or in separate stellar organizations, such as globular clusters and the Magellanic clouds" was reasonable since no evidence available at that time would have lead anyone to a contrary conclusion. Shapley kept it simple, but he got it wrong all the same. In this case, the answer was more complicated.

Discoveries by Walter Baade, Enemy Alien

Both Type I and Type II Cepheids do exist. Consequently, Shapley's period-luminosity relationship was calibrated incorrectly and Hubble's distances to extragalactic nebulae were wrong. This discovery emerged when Walter Baade recognized that spiral galaxies have two distinct populations of stars. Baade was a German-born astronomer who had been working at Mount Wilson Observatory since 1931. In the midst of the Second World War, he, like many other German and Japanese immigrants, was declared an enemy alien. Unlike many enemy aliens

Figure 20.1. Walter Baade. Image courtesy of the Huntington Library, San Marino, California.

of Japanese descent, Germans were not sent to internment camps; in April 1942, however, military security in Los Angeles restricted Baade to his home between the hours of 8PM and 6AM. This effectively put a stop to his work as an astronomer.

To Walter Adams, then the director of Mount Wilson Observatory, this situation was unacceptable. Adams convinced the local army command to make an exception for Baade, which permitted Baade to return to Mount Wilson for nighttime observing sessions. In the summer of 1942, after Hubble left Mount Wilson to work at Aberdeen Proving Ground in Maryland for the duration of the war, Baade took over the observing time that otherwise would have been Hubble's. Baade was given lemons (World War II; enemy-alien status; nighttime, partial blackouts in Los Angeles) but he promptly used them to make lemonade. Long nights with dark skies were ideal for astronomical research; Baade had access to the largest telescope on the planet; and thanks to Hubble's wartime service he had plenty of time to use it. By 1944, through his studies of Andromeda, Andromeda's two companion galaxies M32 and NGC 205, and a handful of other nearby galaxies,

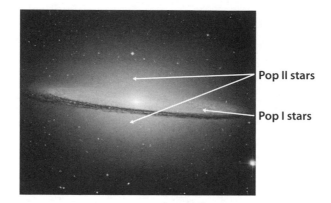

Pop II stars

Pop I stars

Figure 20.2. The Sombrero Galaxy (M104). Walter Baade discovered that spiral galaxies have two distinct populations of stars: Pop I stars are in the galactic disk and include many young blue stars, while Pop II stars are exclusively old red stars in the galactic halo. Image courtesy of NASA and the Hubble Heritage team (STScI/AURA).

Baade had discovered that spiral galaxies like the Milky Way contain two distinct populations of stars. His "Population I" stars included the hot and luminous O and B stars and all the stars in open clusters and in spiral arms. "Population II" included all globular cluster stars, including Bailey's short-period (few hours) cluster-type variables (Chapter Fifteen).

In 1948, with the commissioning of the new biggest telescope on Earth, the 200-inch-diameter Hale Telescope at Palomar Mountain, Baade announced that one of the first projects for this instrument would be to solve the Cepheid problem once and for all. Within a few years, he had established the existence of two distinct populations of Cepheids, each with a unique period-luminosity relationship. He found Type I Cepheids among the Population I stars and the Type II Cepheids among the Population II stars. While both types of Cepheids pulsate in similar ways, Type Is are younger, more massive stars than Type IIs. Shapley, in 1918, did not have enough or good enough data to recognize that he had mixed different kinds of Cepheids together into a single period-luminosity relationship. After Baade's work, however, astronomers knew that a Cepheid with a pulsation period of about ten days could have a luminosity either about 5,000 times greater than that of the Sun (a Type I Cepheid) or about 1,300 times greater than that of the Sun (a Type II Cepheid).

The sample Shapley used in calibrating his period-luminosity relationship was dominated by the fainter, Type II Cepheids, so he had effectively found the period-luminosity for these stars. The Cepheids that Hubble found in Andromeda and other extragalactic nebulae, however, were of the intrinsically brighter Type I. Consequently, Hubble underestimated the brightnesses of the Cepheids in Andromeda by a factor of four and so underestimated their distances by a factor of two. His measuring stick had not been calibrated correctly.

In 1952, in a report he made to the International Astronomical Union in Rome, Baade proposed that the Andromeda Galaxy was nearly three times more distant than Hubble had believed in 1929. (In 1930, Robert Trumpler showed that interstellar dust absorbs light from distant objects, making them appear fainter and thus closer than they actually are; interstellar dust and the recalibration of the period-luminosity diagram both contributed to Baade's estimate of the greater distance to Andromeda). Suddenly, the universe was much bigger and, by virtue of being bigger, much older. In 1956, Edwin Hubble's protégé Allan Sandage improved and corrected Hubble's work and announced that the value of H_0 was 75 kilometers per second per megaparsec, and that therefore the Hubble time was 13 billion years, with the value of H_0 uncertain by a factor of two (thus, H_0 might be 35 or 150 kilometers per second per megaparsec; and the Hubble time might be 30 or 7 billion years). Whether it was 7 or 13 or 30 billion years, the Hubble time as now estimated was long enough to be consistent with the age of the Earth as determined from the radioactive dating of rocks.

During the three decades that followed, measurements of H_0 ranged from about 50 to about 100 kilometers per second per megaparsec, with the two most active and formidable research groups in the field championing either one or the other of the two extreme values, either $H_0 = 50$ kilometers per second per megaparsec (advocated by Sandage, among others) or $H_0 = 100$ kilometers per second per megaparsec. The value $H_0 = 50$ kilometers per second per megaparsec was known as the "long" value and $H_0 = 100$ kilometers per second per megaparsec as the "short" value, since for a given measured redshift velocity, the distance

Figure 20.3. Edge-on spiral galaxy NGC 4013. In 1930, Robert Trumpler proved that interstellar dust exists and both dims and reddens starlight from distant objects. The dark lanes in spiral galaxies are due to the presence of enormous amounts of dust between the stars. Image courtesy of NASA and the Hubble Heritage team (STScI/AURA).

to an object will be longer if the value of H_0 is smaller and will be shorter if H_0 is larger. In 1986, Michael Rowan-Robinson, in his book *The Cosmological Distance Ladder*, critically assessed half a century of measurements of H_0 and concluded that the best estimate for the value of H_0 was 67 ± 15 kilometers per second per megaparsec. This value of H_0 placed the age of the universe in the vicinity of 15 billion years or, more conservatively, in the range from 12 to 20 billion years.

The Hubble Space Telescope and the Age of the Universe

In April 1990, after more than a decade of design and development, NASA launched the Hubble Space Telescope. One of the Key Projects was to make a definitive measurement of the Hubble constant. Along with twelve other members of their team, co-leaders Wendy L. Freedman, Robert C. Kennicutt, Jr., and Jeremy R. Mould focused on identifying Cepheid variable stars in galaxies at distances out as far as 25 megaparsecs (80 million light-years). They reported their final results in 2001: H_0 = 72 ± 8 kilometers per second per megaparsec.

Since the close of work on the Key Project, Adam Riess and his collaborators have continued this work as part of their SHOES project, also done with the Hubble Space Telescope, by identifying and measuring the periods of 240 Cepheids in galaxies as far away as about 30 megaparsecs (100 million light-years) and in galaxies whose distances are well known from observations of Type Ia supernovae (Chapter Twenty-one). Riess' results, reported in 2009, refine the value of H_0 even further, yielding the result H_0 = 74.2 ± 3.6 kilometers per second per megaparsec.

More than half a century of hard work by hundreds of astronomers has improved the accuracy of our H_0 value from Sandage's "a factor of two" uncertainty to an impressively small margin of error of less than 5 percent, but the value of H_0 as determined by the Hubble Key Project team and SHOES team is almost identical to that found by Sandage in 1956. Ironically, Sandage and his collaborators continue to maintain that the Key Project result is wrong and that the actual value of H_0 is close to 62 kilometers per second per megaparsec.

With a reliable value of H_0 in hand, our best estimate for the Hubble time, that is the age of the universe assuming a Hubble constant unchanged in value over the entire history of the universe, is about 13.5 billion years (we will see, from the discoveries described in the next five chapters, that the best estimate for the age of the universe is about 96 percent of the Hubble time). We should note that this age is the result of making measurements and that all measured values are imperfectly known (hence the ± 3.6 notation on the value of the Hubble constant). As a result, scientists typically give a range of values for the answers they calculate, which encompasses the most reasonable possible values. In this case, cosmologists are nearly 100 percent certain that the Hubble time must be greater than 12 and less than 16 billion years.

··· CHAPTER 21 ···

The Accelerating Universe

Its energy surrounds us and binds us. Luminous beings are we, not this crude matter. You must feel the Force around you; here, between you, me, the tree, the rock, everywhere, yes. . . . Yes, a Jedi's strength flows from the Force. But beware of the dark side.

—Yoda, in *Star Wars Episode V: The Empire Strikes Back* (1980)

For decades, teams of astronomers competed to make the most precise measurements of the current value of the Hubble constant. Of course, since the universe is expanding we know that the Hubble "constant" is a misnomer: it should not have been constant over the entire age of the universe, and this is why it is best to identify it as the *current* value of the Hubble constant. To make this point clear in their mathematics, astronomers use the notation H, without the subscripted zero, to denote a Hubble constant that changes with time and H_0 to indicate the value of the Hubble constant during the current epoch in the history of the universe. Now, remember that the Hubble constant measures the rate at which galaxies rush away from each other as part of the cosmic expansion of space—as measured by us today in our part of the universe—divided by the separation distance between the galaxies. If the rates of separation remain constant, the Hubble constant decreases because the separation distances increase. In a universe with mass, gravity acts to slow down the rate of expansion. As the rate at which the universe expands slows down, the value of the Hubble constant should decrease even faster than if gravity played no role. Based on only these considerations, it is clear that as the universe ages H should get smaller.

The Hubble "Constant" Is Not Constant

Let's try this one more time: Assuming that gravity is the dominant force affecting the expansion rate of the universe, the universe should have been expanding faster in the past. If the expansion rate was larger and the universe was smaller in the distant past, the Hubble constant (expansion rate divided by separation distance) should have been larger then than it is today. And where does an astronomer look to find information about "the past"? As we have seen, information about the present is found in the nearby parts of the universe, but information about the past is far away. In the messages conveyed by light from the distant universe (remember, we are looking into the past), at large cosmological redshifts, the Hubble constant should have a larger value than in the local universe. What we need, then, in order to measure the expansion rate of the universe and determine if it has changed, are standard candles that can be measured out to great cosmological distances. For the Hubble Key Project and the SHOES project, Freedman and Riess and their collaborators measured the value of H for Cepheids out to distances of about 30 megaparsecs (100 million light-years), which corresponds to a look-back time of only about 100 million years. 100 million light-years is close enough that cosmologists consider the volume of space out to that distance to be within our "local" universe. What astronomers needed was the ability to look outwards to a distance of 1 billion light-years or more. When we achieved that goal, astronomers predicted, we would find that the value of H had in the distant past been larger than the current-day value of 74 kilometers per second per megaparsec.

The Chandrasekhar Limit

How do we measure the Hubble constant in the distant universe? All we have to do is find a standard candle that we can identify at distances of hundreds of millions to billions of light-years. Since more distant objects must be extremely bright in order for us to see them, we need to observe the brightest possible objects that we recognize as standard candles.

Remember our white dwarfs? We noted that an isolated white dwarf, when left alone, maintains a constant mass and radius and simply cools off slowly in a way that we understand. In doing so, cooling white dwarfs provide us with a tool for measuring their ages and the ages of the ancient globular clusters in which they are found. Not all white dwarfs are isolated, however, and in a binary star system a white dwarf can grow in mass at the expense of its companion. If we have a binary system that contains both a white dwarf and a red giant, mass can flow from the outer atmosphere of the red giant onto the surface of the white dwarf. Later, during the planetary nebula phase, the red giant will expel mass into space, and some of that mass also will fall onto the surface of its white-dwarf companion. As a result, the white dwarf will slowly grow.

Normal objects, for example a pile of sand, become bigger when more mass is added to them. White dwarfs, however, do the opposite. More mass means a stronger gravitational force to compress the white dwarf. Consequently, adding mass to a white dwarf makes it smaller. Greater compressive force means that both the internal pressure and the density of the white dwarf increase. When the pressure and density become high enough, the inward force of gravity overwhelms the outward pressure exerted by the degenerate electron gas. It is as if, in our game of musical chairs, the chairs are pushed so close together that their legs become entangled. When someone sits on the tangled chairs, the chair legs twist and shatter. The entire edifice collapses. In our white dwarf, too much mass overwhelms the electron degeneracy pressure that was resisting gravity and providing the pressure to maintain the structure of the star. Triggered by its mass becoming just a smidgen *too* massive, the white dwarf collapses in on itself. The critical threshold for a "too massive" white dwarf is about 1.4 solar masses and is known as the *Chandrasekhar limit*.

Type Ia Supernovae

White dwarfs do not collapse in on themselves when their masses grow to 1.0 or 1.2 or 1.3 or 1.35 or 1.45 or 1.7 or 5 solar masses. Electron degeneracy pressure fails at the precise limiting pressure that is exerted by a specific known mass. In every binary system in which a white dwarf accretes mass from a red giant companion and reaches the Chandrasekhar limit of 40 percent more mass than the Sun, the white dwarf will collapse; because all white dwarfs collapse at the Chandrasekhar limit, the ways in which they collapse and then explode as supernovae will be extremely similar, and so they will all be nearly identical in brightness.

Inside the collapsing white dwarf, the central density becomes great enough that carbon fusion begins. Over a few centuries, the release of heat from these nuclear fusion reactions thaws the cold white dwarf. When the rising temperature reaches about 700 million K, large-scale ignition of thermonuclear reactions spreads to multiple locations inside the white dwarf. Within a fraction of a second of large-scale ignition, the nuclear reactions consume virtually the entire white dwarf. Carbon fuses to neon, magnesium, and sodium; oxygen fuses to sulfur and silicon; and silicon fuses to nickel. This rapid sequence of nuclear reactions almost instantly turns 40 to 60 percent of the white dwarf into nickel. Immediately, the nickel begins to decay to cobalt and iron. The iron nuclei in the core absorb photons, causing the iron nuclei to disintegrate into alpha particles (helium nuclei) and the core to cool off, with catastrophic consequences for the white dwarf. The overlying layers of the star continue to crush the core. As a result the alpha particles dissociate into protons, neutrons, and electrons. Electron degeneracy pressure cannot halt the collapse; quickly, the electrons are crushed into the protons to form a core of neutrons. Almost instantly, the core is transformed into the equivalent of an atomic nucleus made of only neutrons and with a diameter of only a few kilometers. The neutrons then exert a pressure (neutron degeneracy pressure) that pushes back against the collapsing outer shells of the dying, imploding white dwarf. The outer layers of the white dwarf bounce off the core, sending a shock wave outwards. The gravitational strength of the former white dwarf is typically too little to contain the enormous amount of explosive energy generated during the collapse and rebound, so the white dwarf explodes. Within a few seconds, the rapid drop in density in the expanding, exploding star brings the nuclear fusion reactions to a halt.

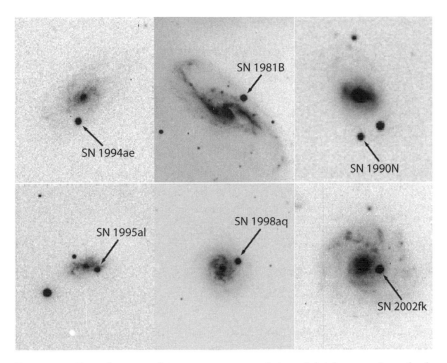

Figure 21.1. Optical images of six supernovae near their peak brightnesses. In each of these images, the supernova is seen when its brightness is comparable to that of its host galaxy. From Riess et al., *The Astrophysical Journal* (2009): 699, 539.

The light emitted by an exploding white dwarf can be as great as a few billion suns and is comparable to the amount of light emitted by an entire galaxy, though it lasts for only a brief time. Astronomers call such an object a *supernova*. From the standpoint of observational astronomy, supernovae are fantastic objects, sometimes as much as 100,000 times brighter than the brightest Cepheid variable stars. Their blaze of glory, however, is short-lived; as seen in visible light supernovae fade away into obscurity in only a few months. They fade away because the star is gone, literally blown apart in the explosion. Most of the light we see comes from the decay of the enormous amounts of radioactive nickel produced during the explosion. The decay of nickel to cobalt has a half-life of only six days; the decay of cobalt to iron has a half-life of seventy-seven days. Within a few months and a few half-lives of radioactive cobalt, the supernova fades away and disappears from sight.

Very massive stars also can explode as supernovae at the end of their lifetimes and can do so with masses of 20 or 50 solar masses or with whatever mass the star had when iron nuclei in the core began to absorb photons and disintegrate into helium nuclei. Supernovae from massive stars are also as bright as billions of suns but, unlike white dwarfs, they are not all identical in brightness since their progenitor stars can have had any mass larger than about eight solar masses. For reasons having to do with the chemical signatures observed in the

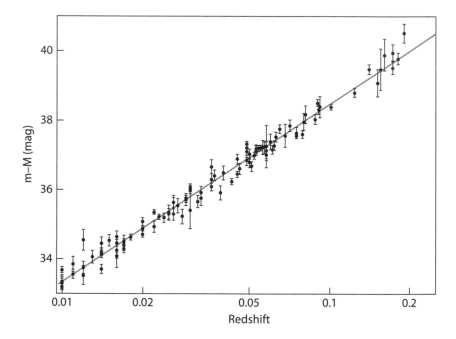

Figure 21.2. Observations of supernovae at relatively low redshift (low z), plotted with redshift versus distance. (The difference between the apparent and absolute magnitudes, plotted on the y-axis, can be thought of as a proxy for distance.) The straight line through all the data points indicates that the inverse square law for light applies for objects out to redshifts of $z = 0.2$, and therefore that the expansion rate of the universe is very constant over the time period measured by these observations. From Perlmutter and Schmidt, "Supernovae and Gamma-Ray Bursters," in *Lecture Notes in Physics*, K. Weiler, ed. (2003): 598, 195. Reproduced by permission of Springer Publishing.

spectra of these supernovae, supernovae from massive stars are known as either Type Ib, Ic, or II. Thus, from their spectra, astronomers can easily and confidently distinguish supernovae that come from the explosions of massive stars from supernovae that come from exploding white dwarfs, which are categorized as Type Ia supernovae.

The crucial detail that makes Type Ia supernovae good standard candles is that they all arise from white dwarfs that explode at nearly identical masses; therefore, they are very similar, though not identical, in brightness. The big step forward in supernova research came when astronomers learned to distinguish, on the basis of their light signatures, which Type Ia's are a bit over luminous and which ones are a bit under luminous. The ability to identify too bright vs. too dim Type Ia supernovae meant that researchers also could identify those of normal brightness. Once they knew the absolute brightness of normal Type Ia supernovae, which they could determine from supernovae in the Milky Way or in nearby galaxies that contain Cepheids, they could use the measured apparent brightness of any newly discovered Type Ia supernova and the absolute magnitude for this standard candle to determine the distance to the new supernova, once they had decided whether it was an over bright, under bright, or just right supernova. Thus, for each new supernova, astronomers can place one more data point on a Hubble diagram. From a survey of such measurements for at least several Type Ia supernovae, we can extend the Hubble diagram out to great distances.

The Supernova Cosmology Project and the High-z Supernova Search

In the 1990s, two independent teams of astronomers began doing exactly this. One group, led by Saul Perlmutter at the Lawrence Berkeley National Laboratory, dubbed itself the Supernova Cosmology Project. The other group, led by Brian Schmidt and based at the Harvard-Smithsonian Center for Astrophysics, took the moniker the High-z Supernova Search (High-z refers to large cosmological redshifts, z being the letter astronomers use for redshift).

It is impossible to predict when a particular white dwarf is going to explode, which is a source of considerable inconvenience for astronomers who study supernovae. We cannot relax in our lounge chairs, sipping martinis while watching and waiting for the scheduled explosion. We do know, however, that astronomers have observed seven supernovae explosions in the Milky Way over the last 2,000 years: SN 185, observed in 185 CE in the constellation Centaurus; SN 393 in Scorpius; SN 1006 in Lupus; the Crab Nebula supernova of 1054; SN 1181 in Cassiopeia; Tycho's Supernova of 1572; and Kepler's Supernova of 1604. Most likely, an equal number of supernovae occurred but were not observed over the same period. Assuming, therefore, that the last two millennia were typical for any two-thousand-year interval in any spiral galaxy, we can estimate that, on average, about one supernova will erupt in a typical spiral galaxy every century. Obviously, this is a very imprecise estimate, but since we have observed seven supernovae in twenty centuries (0.35 per century), we can feel fairly confident that if we had observational data for millions of years rather than for only two thousand, we would find that the supernova rate would be within a factor of a few of 0.35 per century; that is, it would almost certainly be bigger than 0.1 and less than 10 per century. Recent statistical work analyzing data from modern searches for supernovae in nearby galaxies yields the same result: about one supernova per century in a galaxy like the Milky Way.

This information tells us that if we observed one galaxy for a hundred years, we probably would see one supernova explosion when it was within a few days of maximum brightness, which would be our goal. But astronomers, whose careers are typically a bit shorter than one hundred years, need a better strategy for finding supernovae. What if we observed a hundred galaxies for one year? Statistically, we should detect one supernova near maximum brightness from that set of observations. But what if we want to detect one supernova near maximum brightness tonight? If we assume that any one supernova will remain near its maximum brightness for as long as one week, we need to increase our odds by about 50 (approximately the number of weeks in a year), which means that tonight we need to observe about 5,000 galaxies.

What if we're unlucky and we detect the wrong kind of supernova, perhaps a Type II? We could play it safe and observe 50,000 galaxies tonight, in which case we could expect to discover as many as ten supernovae near maximum brightness, of which a few might well be of the type that interest us. From additional observations of these, we would identify the Type Ia supernova that we wanted to study in more detail. Once we had our Type Ia supernova, we would study it further in order to carefully measure its apparent magnitude and redshift. With those numbers in hand, we could calculate the distance to the supernova.

Both the SCP and High-z teams employed strategies not unlike the one we have just outlined. They used a method pioneered in the 1980s by the Danish astronomers Leif Hansen, Hans Ulrik Nørgaard-Nielsen, and Henning Jørgensen for finding supernovae efficiently and then quickly obtaining crucial follow-up data before they faded. It works as follows. First, the teams obtain images of tens of thousands of galaxies. Then they obtain a second set of images of these same galaxies one month later. If, during the intervening month, a supernova has erupted in any of the galaxies in the image sets, the observers can discover the presence of this supernova by comparing the before and after images. These comparisons produce a target list for another observing team that is on stand-by at another telescope, prepared to make follow-up observations that will enable the teams to collect the additional data they need to make the standard candle measurements they are after. In practice, because the second observing team is waiting and telescope time is precious, team members must complete this enormous image comparison task within a few hours. The Danish team was working on a relatively small telescope, which presented a serious disadvantage when trying to observe faint objects in the distant universe; in addition, they did not have the large detectors that enabled the SCP and High-z teams in the 1990s to quickly survey enough galaxies to find a statistically useful number of supernovae. Those detectors were still in development. After two years of effort, the Danes gave up, having reported only one Type Ia supernova detection. Half a decade later, technology caught up with good ideas and the SCP and High-z teams began to detect Type Ia supernovae at a high enough rate to pursue their goal

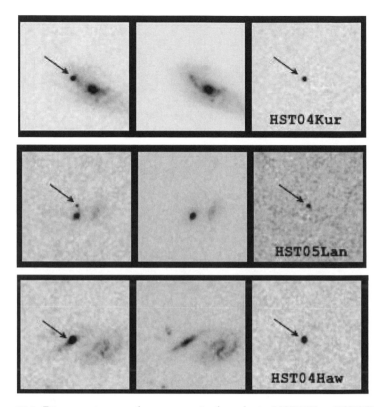

Figure 21.3. Discovery images of supernovae in the galaxies HST04Kur, HST05Lan, and HST04Haw. The left panels show images after the supernovae exploded, the center panels images taken before the supernovae exploded, and the right panels the differences between these pairs of images (each "before" image is subtracted from its partner "after" image). From Riess et al., *The Astrophysical Journal*, (2007): 659, 98, and the High-z team.

of measuring the value of the Hubble constant in the more distant universe.

What should we expect from measurements of the redshifts and brightnesses of distant supernovae? Imagine that gravity does not act to slow down the speeds at which galaxies are rushing away from each other—a scenario that cosmologists call a "coasting universe" or an "empty (no mass) universe." In such a universe, we could use the recession values of nearby galaxies, or equivalently the Hubble con-

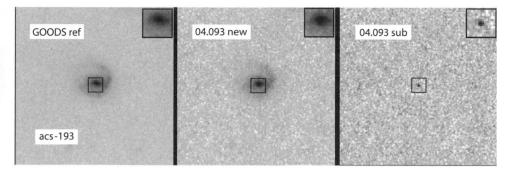

Figure 21.4. Discovery images of a supernova in the galaxy acs-193. The left panel shows an image of acs-193 taken before the supernova exploded, the center panel an image taken after the supernova exploded, and the right panel the difference between these two images (the "before" image subtracted from the "after" image). In this case, the supernova exploded very nearly at the center of the galaxy, making it very difficult to identify the presence of the supernova except in the third image. Images courtesy of Robert Knop.

stant measured from nearby galaxies, to calculate the distances to much more distant objects. The distances to these objects would allow us to predict their brightnesses, and we could compare their predicted to their measured brightnesses. If the predicted and measured brightnesses were the same, we would know that we indeed live in a coasting universe.

What if the universe has mass (which it does) and that mass acts to slow down the rate of expansion of the universe over cosmic time (which it must)? We would call this a "decelerating universe." In it, distant galaxies would be closer to us than they would be if the expansion rate had not decreased at all. If they were closer, they would be brighter than we would predict them to be, under the assumptions for a coasting universe. If the universe contains only a small amount of mass and energy, the deceleration rate of the universe will be small and so the universe will have expanded at nearly the same velocity for all of universal history. On the other hand, if the universe contains a large amount of mass and energy, the rate at which the universe has expanded should have decreased significantly over billions of years. In

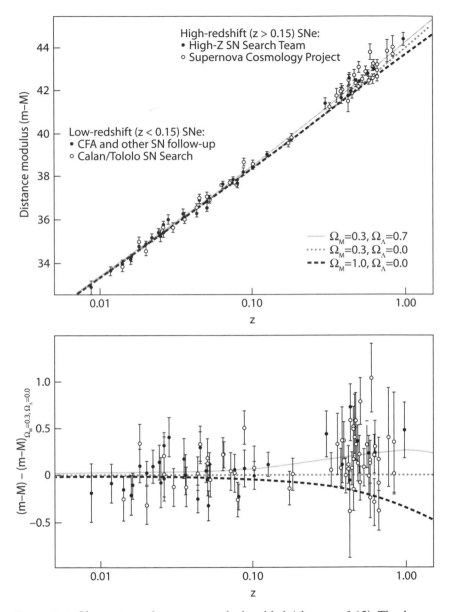

Figure 21.5. Observations of supernovae at high redshift (above $z = 0.15$). The data from the previous plot fits onto the left side of this plot. The thick dashed line marks where the supernovae were expected to be found on this plot: given Hubble's law for smaller redshifts, a given redshift predicts the distance of any supernova. Top panel,

either case, we would find that distant supernovae would be closer and *brighter* in a decelerating universe than they would be in a coasting universe. This result is what the SCP and High-z teams expected to find.

In 1998, both teams reached the same conclusion: at large distances (large redshifts or high z), the Type Ia supernovae are about 25 percent *fainter* than expected, meaning that they are *more distant than expected*. The results of both the SCP team and the High-z Supernova Search team were the same: the brightnesses of the distant supernovae were outside of the plausible range for a coasting or a decelerating universe. Instead, they were too faint and therefore too distant.

The only logical explanation for these distant supernovae being as far away as they are is that the universal expansion rate is *increasing*. The supernovae have been pushed away from us faster than gravity can pull them back. Some force, or something that acts like a force, is opposing gravity and winning. (More properly, of course, we should say that something is acting to make space expand more rapidly now than in the past, since the motion of distant galaxies is only apparent, caused by the expansion of space and not by the movement of galaxies through space.) Instead of living in a decelerating universe, we live in an accelerating universe: as more time passes, distant galaxies are pushed away from each other faster and faster. The universe, it would seem, has the ability to make space expand faster and faster and can overcome the tendency of gravity, an attractive force, to pull space back together or at least to slow down its expansion.

Figure 21.5. (*Continued*) the supernovae at high redshifts (redshifts are plotted along the horizontal axis) have distances (plotted using a proxy for distance, called the *distance modulus*, along the vertical axis) that make them fainter and farther away (higher on the plot) than expected. Bottom panel, the vertical axis is now the *difference* between the observed distance and the expected distance in a universe in which gravity is slowly and steadily slowing down the rate at which the universe is expanding. The data at the right side in the bottom panel clearly demonstrate that the most distant supernovae are much fainter (positive values for the *difference*) than expected in such a universe. Data is shown from supernovae detected by both the Supernova Cosmology Project team and the High-z team. From Perlmutter and Schmidt, "Supernovae and Gamma-Ray Bursters," in *Lecture Notes in Physics*, K. Weiler, ed. (2003): 598, 195. Reproduced by permission of Springer Publishing.

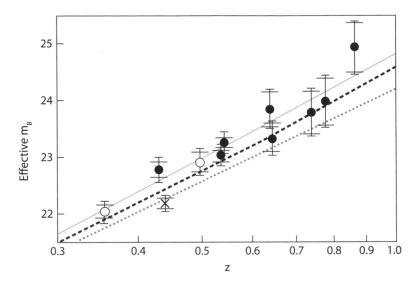

Figure 21.6. Brightness versus redshift plot for supernovae with redshifts from 0.3 to 0.9 discovered with the Hubble Space Telescope by the Supernova Cosmology Project team. The lower (dotted) line shows where the supernovae would be in a coasting universe. The solid (top) line is the best fit to these data. The actual supernovae measurements show that they are fainter (the effective apparent magnitude) than expected at every redshift. Image from Knop et al., *The Astrophysical Journal* (2003): 598, 102.

The Accelerating Universe and the Age of the Universe

We already understood when we calculated an age from the Hubble constant, the Hubble age, that we were assuming that the rate at which the universe had been expanding had never changed. Now we know that this assumption is wrong. The universe is younger than the age we calculated by assuming a constant value of H. Since the currently measured value of H is about 74 kilometers per second per megaparsec, and since this value of the Hubble constant yields an age for the universe of about 14 billion years, we know that the age of the universe must be less than 14 billion years.

If the current, accelerating phase of the history of the universe has been going on for a long time, the universe could be significantly younger than 14 billion years; however, if the accelerating phase of universal history has had only a small effect on changing the value of the Hubble constant, the universe might be very nearly 14 billion years old. All the evidence we have from other constraints, including the age of the universe as estimated from cooling white dwarfs and the turn-off point ages of globular clusters, suggests that an age in the vicinity of 13.5 to 14 billion years is most likely.

Einstein's Cosmological Constant

Albert Einstein had postulated, at least in mathematical terms through the addition of his "cosmological constant" to his equations for general relativity, the existence of a force that would oppose gravity. He had done so because he thought a repulsive force was needed in order to preserve a static universe, to prevent the universe from collapsing back on itself. Einstein's cosmological constant represented that repulsive force, that negative pressure that opposed gravity. After Hubble discovered that the universe was expanding, Einstein discarded the cosmological constant from his own equations. Now, three quarters of a century later, it begins to look like yet another brilliant insight on his part.

Dark Energy

Physicists and astrophysicists do not yet know what provides this anti-gravity force, though they know that it is only manifest at vast, cosmological distances. So far as we can tell, it is not important in our daily lives and has no impact on objects in our solar system, or even in our galaxy. But although we don't know what the **stuff** is, it now has a name: *dark energy*.

The leading suggestion is that dark energy is the vacuum energy density of space. According to quantum mechanics, what we might think of as empty space is never quite empty. It is filled with particles and anti-particles that are constantly being created and annihilated. As long as a particle and an anti-particle are not around for very long— they are called virtual particles in this kind of situation—they follow all the so-called "conservation rules" of physics, including the best- known of them; namely, "the law of conservation of energy." The con- tinual creation and annihilation of these particles and anti-particles produces a pressure in space that makes space expand. Across short distances in the universe, gravity is dominant. Across very large, cos- mological distances, all those little burps of pressure add together to generate enough expansive pressure to dominate gravity.

Science magazine and the American Academy for the Advancement of Science identified the discovery of the accelerating universe as the most important scientific discovery of 1998. Since that pivotal year, dark energy, whether it is a newly discovered kind of energy or a fifth force, has quickly emerged as one of the principal research topics of physicists and astronomers. Two questions dominate their investiga- tions. What is dark energy? And how much of the total energy content of the universe is contained in dark energy? The answer to the first question is of great interest as it impacts the fundamental nature of physics and of the universe. The answer to the second question, how- ever, is of greater interest to us, as the fractional content of the uni- verse that is dark energy is going to be vital to our last strategy for measuring the age of the universe (Chapter Twenty-six).

The nature of dark energy is but one of two closely related mysteries in modern astrophysics. The other concerns dark matter, for which we find ourselves asking the same two questions. What is it? And how much of the total energy content of the universe is contained within it? Our answers about dark matter, like those concerning dark energy, also will impact our final measurements of the age of the universe.

···CHAPTER 22···

Dark Matter

For R > 8.5 kpc, the stellar curve is flat, and does not decrease as expected for Keplerian orbits.

—Vera Rubin, in "Kinematic Studies of Early Type Stars. I. Photometric Survey, Space Motions, and Comparison with Radio Observations," *The Astronomical Journal* (1962)

The concept of dark matter is one of the most exciting, compelling, mysterious, and perhaps disturbing ideas that has emerged from modern astrophysics. Besides being intriguing in its own right, dark matter has enormous implications for our understanding of how we fit into the universe. And it turns out to be crucial when we use the cosmic microwave background (Chapter Twenty-four) to determine the age of the universe (Chapter Twenty-six).

The term *dark matter* means, quite simply, matter that produces so little light that we cannot see it. We can, however, detect the presence of dark matter even when it does not generate much or any light. But dark matter is not simply normal matter that generates very little light; if it were, it would not be such a powerful and bizarre concept. Instead, in the twenty-first century, we know of several kinds of dark matter, some familiar, others decidedly exotic.

Some kinds of dark matter are objects composed, ultimately, of subatomic particles that we (or at least physicists) understand from laboratory experiments and have become familiar with, if only by name: protons, neutrons, mesons, and a host of other particles composed of quarks, as well as electrons, muons, tauons, and neutrinos. Depending on the combinations of quarks that make up a particle, those particles can respond to some or all four forces: the strong nuclear force, the weak nuclear force, the electromagnetic force, and the gravitational force. The other kinds of subatomic particles, that is, those not composed of quarks, do not respond to the strong nuclear force, and the neutrinos only respond to the weak nuclear force and gravity.

Exotic dark matter, on the other hand, responds to the force of gravity but, depending on what kind of exotic dark matter it is, does not respond to one or more of the other forces. For example, with regard to the electromagnetic force, the most exotic forms of dark matter particles do not emit, produce, or interact with light in any way. Examples of such exotic dark matter particles are axions and WIMPS (Chapter Twenty-three). Such matter is truly, completely dark. The existence of exotic dark matter has enormous implications for helping us identify how we fit into the universe and ultimately for determining the age of the universe.

Tracing the progress of astronomers' ideas about dark matter over the course of the last two and a half centuries will prove enlightening, even though some of the first-identified forms of dark matter are no longer considered to be such. This historical look will help us understand the evidence that convinced astronomers over time that dark matter, especially exotic dark matter, must exist. Our problem can be reduced to this question: How do we know something is out there in the heavens if we cannot see it? After all, as we are fond of saying, seeing is believing.

The dark matter story began in 1783, when British geologist and amateur astronomer, the Reverend John Michell, imagined the possible existence of a star 500 times bigger in radius than the Sun but with the same average density as the Sun. He calculated that "all light emitted from such a body would be made to return toward it, by its own proper gravity." He elaborated: "If there really should exist in nature [such] bodies . . . since their light could not arrive at us, we could have no information from light. Yet if any other luminous bodies should happen to revolve about them, we might still perhaps from the motions of these revolving bodies infer the existence of the central ones." In simpler terms, such a star would be dark. It would contain matter that we could not see because no light emitted by this star could ever reach us; however, we could readily infer the existence of the dark star from its gravitational influence on other objects.

Reverend Michell's dark bodies do exist. We call them *black holes*, and they come in several varieties. We know that some black holes form from normal matter, for example those that come into being when some massive stars die (Chapter Twenty-one); others may have formed from exotic dark matter. But since we cannot see inside a black hole, we have no way of knowing anything about the character of the mass out of which any particular black hole is made.

Half a century later, Friedrich Wilhelm Bessel suggested that the stars Procyon and Sirius slalom slowly across the sky in a way that only makes sense if they are each orbiting a dark, unseen companion. Anticipating our modern view of the universe, Bessel wrote, "We have no reason to suppose that luminosity is a necessary property of cosmic bodies." When Alvan Clark detected Sirius B in 1862, these previously invisible companions changed in our understanding from being dark and unseen to being merely faint and hard to detect. They are now known as white dwarfs, and we do not consider the white dwarf com-

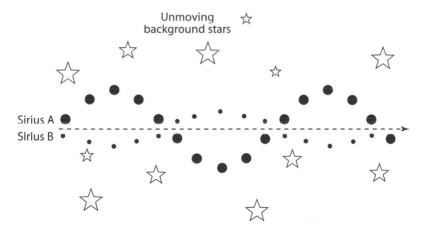

Figure 22.1. Illustration of the motion of a binary star system (here Sirius A and B) as they slalom past the fixed stars over a time period that spans many decades. For the first twenty years after the curved-path motion of Sirius A, with respect to the background stars, was first detected, even the most powerful telescopes were not strong enough to permit observers to detect the presence of Sirius B.

panions of nearby stars to be dark matter. At greater distances, however, white dwarfs are impossible to detect via the light they emit because they are so faint; distant white dwarfs, though they emit light, therefore remain a dark matter candidate and could be an important component of the dark matter in galaxies.

In 1845, the British mathematician John Couch Adams and the French astronomer Urbain Le Verrier independently used well-known problems with the orbit of Uranus to predict the existence and calculate the orbital path of an unseen planet. One year later, when Johann Galle, working at the Berlin Observatory, looked in the direction predicted by Le Verrier for the location of the mystery planet, he found Neptune. This planet was never truly dark; it was merely faint enough to be undetectable without the aid of a telescope and, as a result, escaped detection until someone with a telescope aimed it in just the right direction. Astronomers do not think a significant amount of mass remains to be found in our solar system in the form of distant faint planets, comets, Kuiper Belt objects, or even a distant companion star. However, what if the galaxy is populated by a large number of Neptune-like (or even Jupiter-like) giant planets that do not orbit stars? Such objects could be an important component of the dark matter in galaxies, and it is likely that they will remain all but undetectable in coming decades.

Sir James Jeans suggested, in 1922, that since bright stars are easy to see and faint stars are much harder to see, many stars might be so dim that they will continue to escape detection by astronomers with even the largest telescopes. In fact, he estimated that there likely exist "about three dark stars in the universe for every bright star." Even as astronomers build bigger telescopes and detector systems sensitive to smaller amounts of light, the faintest stars do continue to elude our best efforts to find them. Such dark stars have two things in common with Michell's black holes, Bessel's white dwarfs, and Adams's and Le Verrier's unseen planet: they all are in physical environments in which they either emit or reflect very little or no light; and most likely they are all made of normal matter.

White dwarfs emit very little light because they are small in comparison to normal stars; planets emit very little light because they are very small and cool; dark stars are dark either because they are small, are small and cool, or are very far away; black holes emit no light because they prevent light from escaping. In all these cases, we are considering hard-to-see objects that do interact with light. They are forms of dark matter made of or from normal matter. Taken together, they stand for two centuries of work wrapped around a single, very important concept: matter can be present in the universe whose existence we can infer from the behavior of other objects but whose presence cannot be discerned in light emitted from or reflected by that matter itself because these objects emit or reflect very little or no light.

Cold Dark Matter

Fritz Zwicky was an astronomer whose work at Cal Tech, in Pasadena, California, spanned six decades, from the 1920s into the 1970s. In 1933, Zwicky measured the velocities of galaxies in the Coma Cluster and found that they were moving extremely fast, up to several thousand kilometers per second. According to his calculations, these galaxies had velocities so large that, over a time period much less than the age of the universe, most of the Coma Cluster galaxies should have run away, escaped from the cluster. Quite simply stated, the Coma Cluster should not exist anymore, But it does. Zwicky wondered why.

Zwicky's calculations were based on his observations of all the luminous matter—namely, the galaxies—in the cluster. Of course, in 1933 "luminous" meant producing light our eyes can see. The age of gamma ray astronomy, X-ray astronomy, ultraviolet light astronomy, infrared astronomy, millimeter wave astronomy, and radio astronomy was some years off in the future. Using the luminosities of these galaxies in visible light, Zwicky estimated the total mass of the Coma Cluster; employing this mass estimate, he calculated the escape velocity for galaxies within the cluster, that is, how fast a galaxy would have to be moving to overcome gravity and escape from the cluster. A sufficiently fast-moving galaxy, he decided based on his calculations, should be able to escape, while a slower-moving galaxy would remain part of the cluster. But even galaxies moving much faster than the escape velocity

Figure 22.2. The Coma Cluster, a prototypical dense cluster of galaxies. This was the cluster studied by Fritz Zwicky in 1933. Image courtesy of the Sloan Digital Sky Survey Collaboration, www.sdss.org.

he had calculated for the Coma Cluster had not broken away. Zwicky suggested that the Coma Cluster must contain an enormous amount of unseen matter and that therefore the true escape velocity must be much greater than the value he derived from his measurements of the total amount of luminous matter. By his estimate, the Coma Cluster would have to contain 400 times more mass, all of it unseen, than his estimate for the mass of all the luminous matter, in order to provide the gravitational pull that would bind even the fastest moving galaxies to the cluster forever. Modern improvements on and corrections to Zwicky's work have reduced the requirement for unseen matter to a factor of 50 rather than 400, but his conclusion remains solid: as much as 98 percent of the mass of the Coma Cluster is not producing enough

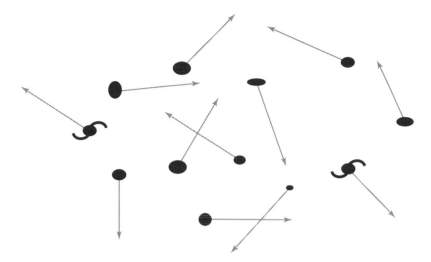

Figure 22.3. Illustration of the motions of galaxies in the Coma Cluster. Zwicky found that their velocities were so high that the galaxies should escape from the cluster.

visible light to be seen. What could that mass be? Black holes? White dwarfs? Jupiters?

Because this unseen matter, like the galaxies, had not escaped the Coma Cluster, Zwicky realized that whatever it was, it must be composed of objects that have velocities comparable to those of the galaxies themselves and much smaller than the speed of light. In an astrophysical environment, slow-moving particles or objects—and the adjective "slow" here means slow in comparison to the speed of light—are called "cold" and fast-moving particles are called "hot." Since this unseen material was "cold" in the logic of astrophysics, Zwicky called it *dunkle (kalte) Materie*, or "dark (cold) matter."

Zwicky's study of the Coma Cluster brought cold dark matter into the twentieth-century conversation, and by 1999 astrophysicist Sidney van den Bergh would write that Zwicky's "may turn out to have been one of the most profound new insights produced by scientific exploration during the twentieth century." Today few would argue with van den Bergh's assessment, but in the 1930s Zwicky's surprising results went largely unnoticed by his colleagues. One who did pay attention was Sinclair Smith of the Mount Wilson Observatory, who in 1936 arrived at a conclusion regarding the Virgo Cluster of galaxies that was strikingly similar to Zwicky's for the Coma Cluster: the large relative motions of thirty individual galaxies, some with velocities of up to 1,500 kilometers per second, would disrupt the cluster unless the average mass for the galaxies was about 200 times larger than the average mass for a galaxy that Hubble had calculated based on the amount of light received from these galaxies (again, modern corrections have reduced the estimate of the unseen mass to a factor of 25 or 30 larger). Smith even speculated, very presciently, that perhaps the mass estimates for the galaxies themselves were correct but that the unseen mass was likely "internebular material, either uniformly distributed or in the form of great clouds of low luminosity surrounding the nebulae."

In 1959, Fritz Kahn and Lodewijk Woltjer, working at the Institute for Advanced Study in Princeton, New Jersey, concluded from a study of what is called our "Local Group" of galaxies, which is a small cluster of galaxies that includes the Milky Way and Andromeda, that "the Local Group of galaxies can be dynamically stable only if it contains an appreciable amount of intergalactic matter." They speculated further,

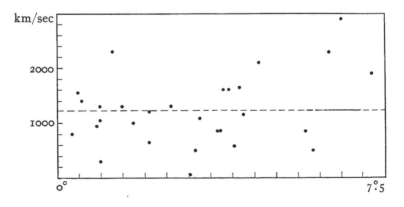

Figure 22.4. The distribution of velocities of thirty galaxies in the Virgo Cluster, as presented by Sinclair Smith in 1936. Some individual galaxies are moving as fast as 1,500 kilometers per second relative to the mean velocity (dashed line) of 1,225 kilometers per second. From Sinclair Smith, *The Astrophysical Journal* (1936): 83, 23. Reproduced by permission of the AAS.

along the same lines as Smith, that most of that invisible mass might be contained in very hot gas filling up the space in between the galaxies. They were right.

Hot Dark Matter

Eventually, some of the matter that Zwicky knew must be present was found, but it was not cold and dark; it turned out instead to be very hot and extremely luminous. Its luminosity was, however, of a kind not detectable with the ordinary tools and telescopes available to astronomers in 1959. X-ray detectors on X-ray optimized telescope would do the job, but they would have to wait for the advent of satellite-based telescopes as the atmosphere of the Earth, fortunately for all the creatures living on it, stops all X-rays from astrophysical sources from penetrating to the surface.

Beginning with measurements made by sounding rockets in the mid-1960s—efforts led by Riccardo Giacconi, who was awarded the 2002 Nobel Prize in Physics for his work in laying the foundations of X-ray astronomy—galaxy clusters stood out as the brightest astrophysical sources of X-rays. With the launch of the first X-ray satellites in the early 1970s, astronomers were able to conclude, with confidence, that the X-rays from galaxy clusters emerge from clouds of hot gas in between the galaxies and not from the galaxies themselves; finally, with the launch of the Einstein Observatory in 1978 and more recently the Chandra X-ray Observatory in 1999, actual images of galaxy clusters taken in X-rays have revealed that the spaces between the galaxies in clusters are filled with gas at temperatures of tens of millions of degrees. At such high temperatures, the intergalactic gas cocooned within a galaxy cluster is a prolific emitter of X-rays.

In contrast to cold dark matter, X-ray-emitting gas is clearly hot, not just in regard to its temperature but in the sense of movement: it travels at speeds of a few thousand kilometers per second. At such high speeds, the gravity of an individual galaxy cannot contain the gas. The total mass of the cluster of galaxies, however, can and does. Now we have cold dark matter and hot dark matter, though this particular kind

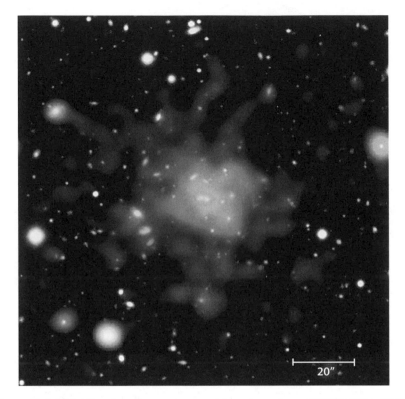

Figure 22.5. This Chandra telescope image of the galaxy cluster RDCS 1252.9-2927 shows the emission of X-rays from 70-million-degree-Celsius gas (seen as the gray and white fuzz and blobs). The X-ray image is superimposed on a visible-light image of the cluster (all of the small dots and smudges are the galaxies seen in visible light). X-ray data indicate that this cluster formed more than 8 billion years ago and has a mass at least 200 trillion times that of the Sun. X-ray image courtesy of NASA/CXC/ESO/P. Rosati et al. Optical image courtesy of: ESO/VLT/P.Rosati et al.

of hot dark matter is not strictly-speaking dark, since it emits copious amounts of X-rays.

The total mass made up of hot, X-ray-emitting atoms that fills the intergalactic spaces within clusters of galaxies is about ten times greater than the total amount of mass that previously had been identified in the form of stars or clouds of gas inside the galaxies themselves. This is an incredible amount of previously unseen matter. No one, prior to the actual discovery of these intragalactic-cluster gas clouds, would ever have guessed that 90 percent of the mass in galaxy clusters exists outside of galaxies in the form of hot gas.

Nevertheless, even this hot, bright X-ray-emitting gas does not constitute enough mass to solve the problem first identified by Zwicky. To solve Zwicky's dark matter problem, we need to find matter that is fifty times more massive than the visible matter in galaxies; the hot intragalactic gas only provides ten times the mass of the visible matter in galaxies. Additional matter, matter three to five times as massive as the total mass in X-ray luminous matter and not detectable in ultraviolet, visible, or infrared light, or in radio waves, or from X-ray telescopic studies, must also be present in these galaxy clusters. Based on some additional evidence, we know that this matter must actually be cold and dark.

Kepler's Third Law, Redux

We need a picture in our mind's eye of spiral galaxies like the Milky Way. Spirals are composed of two main components, a disk and a halo. The luminous disk, which is a flattened structure that contains young and old stars, giant clouds of gas and dust, and large volumes of low-density interstellar medium, extends about 30 kiloparsecs (100,000 light-years) from end to end. Beyond its visible disk of stars are large clouds of hydrogen gas that emit light only at radio wavelengths and can be seen with radio telescopes. Within the disk, the stars and giant clouds generally orbit the galactic center in the same direction. The visible halo, which is composed almost entirely of old stars (Baade's Population II stars), is ellipsoidal in shape, encloses the disk, and is comparable in size to the disk; however, in contrast to the stars and clouds in the disk, halo stars orbit the galactic center in all directions.

With the help of Newton's version of Kepler's third law, which is a direct consequence of the law of gravity, astronomers can use the orbital velocities of stars and gas clouds in the disks of spiral galaxies to learn about the masses of galaxies. First, they measure the velocities of stars or clouds in orbit around the center of a galaxy. For a single star or cloud, the velocity, combined with the distance of the star or cloud

Figure 22.6. Two spiral galaxies. Left, NGC 891. This galaxy is viewed edge on. Right, M101, the Pinwheel galaxy. This image shows how a spiral galaxy looks when viewed from directly above its midplane. Images courtesy of NASA and 2MASS (left), European Space Agency & NASA (right).

from the galactic center, immediately yields the orbital period around the center of the galaxy for that object. The orbital period and the size of the orbit, when combined with Kepler's third law, directly give us the total mass *inside the orbit*.

If a planet were orbiting a double star system, the masses of both central stars would control that planet's orbit. Now imagine a star that is surrounded by a spherical cloud of asteroids. The cloud contains tens of millions of asteroids that, together, have as much mass as the star, and the asteroid cloud has a diameter equal to that of Jupiter's orbit. We put a planet in orbit around this star but a distance ten times greater than the distance of Jupiter, i.e., far outside of the asteroid cloud. In this case, the orbital velocity of the planet would be controlled by the mass of the star plus the mass of the asteroid cloud. The fact that the cloud is neither compact nor at or near the center of the orbit does not matter; the fact that the asteroid cloud is spherical and extends as far above the plane of the planet's orbit as it extends radially outward in the plane of the planet's orbit does not matter. The fact that we do not know the sizes of any of the individual asteroids or even how many asteroids there are in the cloud does not matter.

Now, let's put the same planet in a different orbit. We'll move it well inside the distance of Jupiter, far inside the outer edge of the asteroid cloud. In this location, the orbital velocity of the planet is controlled by the mass of the central star plus the total mass of asteroids inside the orbit of the planet. Even if 50 or 95 percent of the asteroids are in orbits outside of the orbit of our planet, their collected mass does not and cannot affect the orbital velocity of our planet.

Rotation Curves of Galaxies:
Cold Dark Matter

In 1939, Horace Babcock took advantage of these mathematical and physical properties of gravity to measure the orbital velocities of stars in the Andromeda Galaxy. He was attempting to determine how the mass of Andromeda is distributed and to estimate the total mass of the galaxy. Babcock's measurements of the motion of each star in his sample provided an estimate of Andromeda's mass in the volume of the galaxy's space that extended out to the distance of that star. If most of the mass of Andromeda were near the center, then stars at both intermediate and large distances from the center would have orbital velocities determined by the same amount of mass. In such a case, the orbital velocities would decrease as the distances of stars from the center of the galaxy increased, a phenomenon similar to the way in which the orbital velocities of planets decrease with their distance from the Sun. Astronomers call such a pattern of orbits, found in an environment in which virtually all the mass is near the center, "Keplerian."

The Earth, for example, has an orbital velocity of about 30 kilometers per second; the orbital velocity of Jupiter, which is just a bit more than five times further from the Sun than the Earth, is a bit less than two-and-a-half times smaller, at 13 kilometers per second, not the same as Earth's and not five times smaller than Earth's; Neptune is located about 30 times further from the Sun than Earth and has an orbital velocity of about 5.5 kilometers per second, which happens to be about five-and-a-half times smaller. In mathematical terms, when all the mass is concentrated at the center, orbital velocities should decrease in proportion to the square root of the size of the orbit: for an orbit nine times larger than the orbit of the Earth, the orbital velocity would be three times smaller.

On the other hand, if the mass of Andromeda is not concentrated in the center but distributed widely throughout the galaxy, a star at a large distance from the center will be affected by more mass than a star located closer to the center. In this latter case, stellar orbital velocities could increase, stay the same, or decrease slowly as the distances of stars from the center increases, depending on how the mass in the galaxy is distributed.

Measurements of orbital velocities at many distances from the center of those orbits together yield a profile of the distribution of mass in a system. These plots are known as rotation curves. For galaxies, they depict the distances of stars from the center of a galaxy on the x-axis and the orbital velocities of those stars on the y-axis.

What did Babcock discover? Is all the mass in Andromeda near the center, as it is in our solar system, or is the mass distributed uniformly throughout the galaxy? Babcock's results show that the orbital velocities of stars increase from the galactic center out to a distance of about 100 parsecs. These measurements tell us that the mass within 100 parsecs of the center is not concentrated near the center; instead, the amount of mass steadily increases as we move outwards to a distance of 100 parsecs. Moving further out, orbital velocities decrease out to about 300 parsecs. This part of the rotation curve is nearly Keplerian, and suggests that most of the mass within 300 parsecs is actually concentrated within 100 parsecs of the center. Not much mass is added in between 100 parsecs and 300 parsecs. However, from 300 parsecs all the way out to 6 kiloparsecs from the center, the stellar velocities

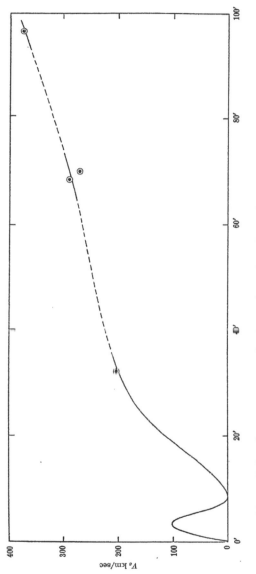

Figure 22.7. Horace Babcock's rotation curve for the Andromeda Galaxy. The vertical axis indicates the velocities of the observed stars, the horizontal axis the distance of those stars from the center of the galaxy (in minutes of arc). Within a few minutes of arc of the center, the orbital velocities increase rapidly with distance, indicating the presence of additional mass with increasing distance from the center. Then the orbital velocities decrease rapidly, indicating that no additional mass is being added □ the total galaxy mass over this distance interval. From about 10′ of arc (600 pc) out to 100′ of arc (6,000 pc), the orbital velocities increase steadily, indicating that a great deal of unseen mass is spread through this part of the galaxy. From Horace Babcock, *Lick Observatory Bulletin* (1939): 498.

steadily increase. This result is only possible if a significant fraction of the mass of Andromeda is distributed throughout the galaxy from distances of 300 parsecs out to 6 kiloparsecs, with vast amounts of mass found at great distances from the center.

The mass of Andromeda had been estimated from the total light emitted by all the stars and clouds of gas and dust observable within the galaxy. The mass distribution inferred from simply looking at where the stars and clouds were located and how much light they produced indicated that a great deal of mass was near the center of the galaxy, with some but not large amounts of mass in the outer parts of the disk. Such a mass distribution should have looked nearly Keplerian, with

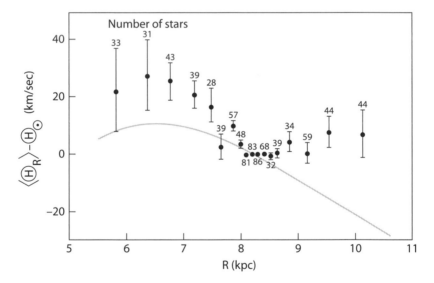

Figure 22.8. Vera Rubin's 1962 rotation curve for the Milky Way Galaxy. The vertical axis indicates the velocities of the stars in comparison to the velocity of the Sun, the horizontal axis the distance from the center of the galaxy (in kiloparsecs). Numbers along the top of the chart indicate the number of stars observed for each data point on the plot. The data show a flat rotation curve from 8 kpc outwards. The solid curve shows the rotation curve for the galaxy based on 1954 measurements of radio signals emitted by Milky Way hydrogen gas clouds within eight kpc of the center, which these data match reasonably well. 1956 models showing what the rotation curve was expected to look like at distances from the center greater than eight kpc clearly do not match these data. From Rubin et al., *The Astronomical Journal*, (1962): 67, 491.

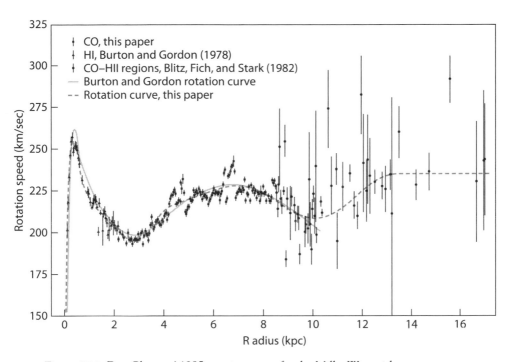

Figure 22.9. Dan Clemens' 1985 rotation curve for the Milky Way, with measurements extending nearly to the center of the galaxy and more than twice as far from the center as the Sun. Clemens' work placed the Sun 8.5 kpc from the galactic center, with a circular rotation speed about the center of 220 km per second. More recent work by Mark Reid and collaborators indicate that the Sun's circular rotation speed is close to 250 km per second. Reid's measurements strongly suggest that the rotation curve of the Milky Way is very similar to that of Andromeda, and so presumably the dark matter components of these two galaxies are also comparable in mass. From Clemens, *The Astrophysical Journal*, (1985): 295, 422.

stellar and cloud velocities decreasing slowly but steadily with increasing distance from the center. This, of course, is not what Babcock found.

If the velocities of the stars are greater than what would be predicted from the amount of mass inferred from measurements of the total amount of light emitted by the galaxy, then additional mass must be present that is unseen and gives no hint of its presence through the emission of any form of electromagnetic radiation. Since this matter

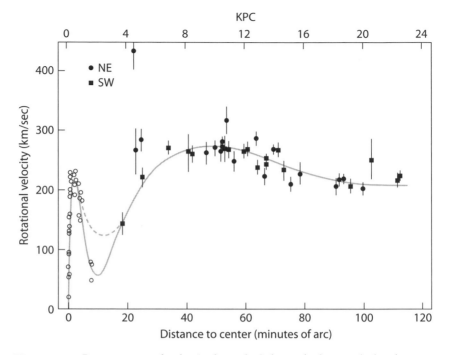

Figure 22.10. Rotation curve for the Andromeda Galaxy, which extends the observed rotational velocities out to 120 minutes of arc, with far more data points than in Babcock's original work. The flat rotation curve continues at least as far out as 7,200 parsecs. From Rubin and Ford, *The Astrophysical Journal* (1970): 159, 379.

has not evaporated from the galaxy, it must be moving at speeds comparable to the speeds of the stars, not comparable to the speed of light, and so it must be cold as well as dark.

In 1962 Vera Rubin took up where Horace Babcock left off. She measured the orbital velocities of 888 stars located within 3 kiloparsecs of the Sun in the Milky Way in order to determine the rotation curve for the galaxy at distances of 5 to 11 kiloparsecs from the center. Rubin's results for the Milky Way were the same as Babcock's for Andromeda. She found that "for R > 8.5 kpc [further from the center of the galaxy than the Sun], the stellar curve is flat, and does not decrease as expected for Keplerian orbits." Such a flat rotation curve can be generated if the amount of mass in the galaxy, at distances from 8 to 11

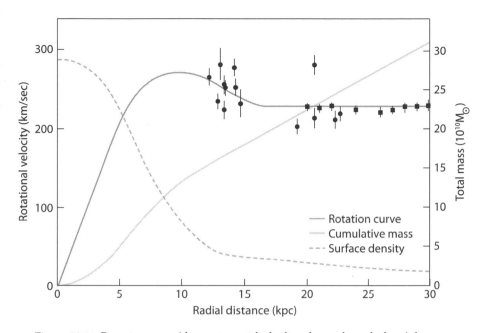

Figure 22.11. Rotation curve (data points with the line drawn through them) for Andromeda that extends the measurements of orbital velocities out to distances of 30,000 parsecs from the center. The line marked "surface density" indicates the amount of mass per square parsec at a given distance from the center. The line marked "cumulative mass" indicates the total mass of the galaxy inside the orbit of a star at a given distance from the center. The flat rotation curve continues at least out to 30 kpc from the center, and the amount of mass climbs steeply from one kpc all the way out to the limits of the data. From Roberts and Whitehurst, *The Astrophysical Journal* (1975): 201, 327.

kiloparsecs, increases with distance from the center at the same rate as the distance increases (mass is directly proportional to distance).

Flat. Another word entered the lexicon of modern astronomy. A "flat rotation curve" for a galaxy indicates that the stars within the galaxy have essentially the same orbital velocities across a vast range of distances from the center. Since the luminous mass in galaxies is insufficient to provide the gravitational force necessary to generate the observed orbital velocities, the only way to generate flat rotation curves is for galaxies to have enormous amounts of mass that is cold and dark.

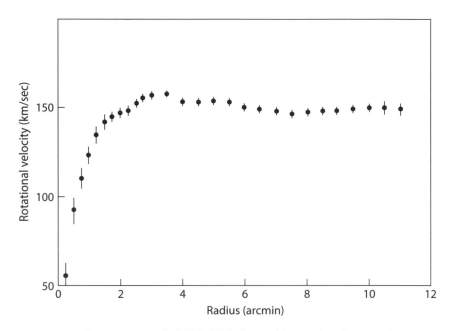

Figure 22.12. Rotation curve for NGC 3198 obtained from radio telescope observations of the orbital velocities of hydrogen gas clouds in this galaxy, showing the flat rotation curve common to galaxies. From Begeman, *Astronomy & Astrophysics* (1989): 223, 47, reproduced with permission © ESO.

In fact, to explain the rotation curve for the Milky Way that Rubin measured, at least 90 percent of the mass of the galaxy must be cold dark matter. Rubin, working in 1970 with Kent Ford, measured the rotation curve for Andromeda using observations of hot, glowing clouds of ionized hydrogen gas and showed that Andromeda has a flat rotation curve all the way out to 24 kiloparsecs from the center; then Morton Roberts and Robert Whitehurst, by studying radio waves from hydrogen atoms in frigid clouds of gas, extended the rotation curve data for Andromeda past the realm of the visible disk of stars, all the way out to 30 kiloparsecs. Still, the rotation curve stayed flat. By the mid-1970s, there was a veritable cottage industry in measuring the rotation curves of spiral galaxies. And without exception, astronomers found the same answer: they were all flat.

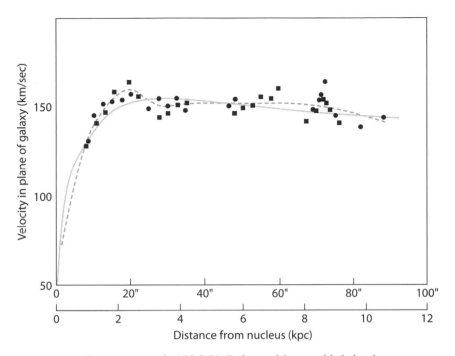

Figure 22.13. Rotation curve for NGC 7217 obtained from visible light observations of orbital velocities of ionized hydrogen gas, showing a nearly flat rotation curve. From Peterson et al., *The Astrophysical Journal* (1978): 226, 770

CHAPTER 23

Exotic Dark Matter

After reviewing all the evidence, it is our opinion that the case for invisible mass in the Universe is very strong and getting stronger.

—Sandra Faber and John Gallagher, in "Masses and Mass-to-Light Ratios of Galaxies," *Annual Review of Astronomy and Astrophysics* (1979)

Astronomers had been finding signs pointing to the existence of dark matter for two hundred years, and by the mid-1980s the evidence was overwhelming. The nature of this invisible matter, however, remained a mystery. Those examples that had been found and identified—black holes, white dwarfs, planets, and hot gas—had turned out to be no different in their physics from the kind of matter that is found on Earth. It was widely believed, therefore, that the still missing matter responsible for the flat rotation curves of galaxies would prove no different.

As we saw in the last chapter, to explain the flat rotation curves of spiral galaxies it is necessary to postulate that some of the mass confined within these galaxies is cold dark matter. Perhaps it consists of great numbers of comets and Jupiter-sized objects floating between the stars. Or perhaps it is dominated by faint red stars or white dwarfs or neutron stars or black holes. None of these choices is likely, however, as observations already made by astronomers place upper limits on how much mass might exist in the form of these kinds of unseen objects. Alternatively, the cold dark matter might be composed of enormous numbers of small subatomic particles, some slow moving (cold), others fast moving (hot), that are hard to find and identify but are nevertheless there, filling up galactic and intergalactic space.

Axions

Axions are hypothetical slow moving, very low-mass particles—each would be one thousand to one million times lighter than an electron—that should have been created abundantly in the Big Bang. They are thought to be extremely stable, with lifetimes much longer than billions of times greater than the current age of the universe. Therefore, any axions spawned in the furnace that existed during the first fraction of a second of the history of our universe (Chapter Twenty-four) should still be around. They are thought to be so abundant, with over 100 trillion of them in every cubic centimeter of space inside the Milky Way and other galaxies, including in the air you breathe, that they should make a large contribution, collectively, to solving the cold dark matter problem. If they actually exist, that is. In our galaxy, axions should fill the galactic halo in which the Sun and Earth are embedded; at any moment, the Earth should be moving through a cloud of axions. All we have to do is detect them, but that is not proving to be so simple.

The very existence of axions, postulated in the mid-1970s when dark matter emerged as an important astrophysical mystery, is central to the theoretical model of physics called quantum chromodynamics (QCD), which explains how the strong nuclear force holds the nuclei of atoms together. Without the axion, the QCD theory would be incomplete, and particle physicists would have to invent another particle with the same properties as the axion in order to make the theory work. Either that or we would have to recraft our ideas about physics so that we had a theory for the strong nuclear force that did not require such particles.

Axions carry no electric charge and can but are nearly incapable of interacting with other kinds of matter through either the strong or weak nuclear forces; that is, they are effectively, but not completely, collisionless. This makes it very difficult to detect their presence. Experimenters believe, however, that despite their stability, they can be tricked: an axion can be induced to decay, to fall apart into other particles, in the presence of a very strong magnetic field. This happens when the axions collide, as improbable as such events should be, with particles of light (photons) that are part of the magnetic field. It should then be possible to detect the axions indirectly by detecting the one or two photons that are the products of that decay event. An experiment at Lawrence Livermore Laboratory in California, the Axion Dark Matter eXperiment (ADMX) uses a magnet that is 200,000 times stronger than the Earth's magnetic field, in an attempt to induce any axions in the environment of that magnet to collide with photons and decay. Other axion searches are underway as well, including one called the CERN Axion Solar Telescope (CAST) that seeks to detect axions that should be created in the core of the Sun when photons collide with electrons and protons. These solar axions should fly right out of the Sun since they are not supposed to be able to easily interact or collide with normal particles in the Sun. Despite the fact that axions must exist to complete QCD theory, neither ADMX nor CAST nor any other axion detection project has yet found any, perhaps because they stubbornly refuse to behave in ways that would permit us to discover their presence, or perhaps they simply do not exist. If axions are not the solution to the dark matter mystery, what other particles or objects might be?

WIMPs

In the late 1970s, the theoretical physics community began positing the existence of special kinds of massive particles, ones with masses tens or hundreds of times greater than the mass of the proton, that likely formed in great abundance early in the history of the universe and that might still be found in the modern universe. These particles were postulated to interact with other matter only through the weak nuclear force and gravity and to be so massive and abundant that they might solve the dark matter problem. Because these Weakly Interacting Massive Particles, or WIMPs, presumably do not interact effectively through the electromagnetic force—they neither emit nor absorb any kind of light—they are impossible to see; and because they respond only weakly, if it all, to the strong nuclear force, they are very difficult to detect in laboratory experiments.

The first objects proposed as WIMP candidates were massive neutrinos (with masses comparable to or greater than that of the electron), which were hypothesized to exist as early as 1977. While massive neutrinos are no longer thought to exist, three kinds of neutrinos (the electron neutrino, the muon neutrino, and the tau neutrino) do exist. Collectively, neutrinos are the second most abundant particle in the universe, out-numbered only by photons, and have properties similar to those of the electron, except that neutrinos lack charge and do not respond to the electromagnetic force. Current generation experiments, however, suggest that the most massive species of neutrino has a mass more than one hundred times less than that of the electron; if so the mass of all the neutrinos in the universe together would be insufficient to provide the amount of dark matter that we know must exist.

Other types of WIMPs have been proposed as part of an effort called *supersymmetry* (SUSY) that aims to merge the theory of quantum mechanics with Einstein's theory of gravity, general relativity. In supersymmetry theories, every normal particle has a *sparticle* (short for "superpartner particle") with all the same properties as the normal particle except the property known as quantum mechanical spin. In addition, sparticles are more massive than their normal-matter partners. Examples of sparticles include "sneutrinos," "selectrons," "squarks," "photinos," "gravitinos," "winos," "zinos," and "neutralinos." Since the least massive sparticles have longer lifetimes than the more massive sparticles, the least massive of all the hypothetical sparticles, the "neutralino," may be the one that still exists in the greatest abundance in the universe. As a result, many physicists believe the neutralino is the best candidate particle for solving the dark matter problem. In the coming decade, the giant particle physics accelerator at CERN, in Geneva, might yield some evidence for the existence of the neutralino, but to date no evidence has been found for its existence or for that of any other sparticles.

MACHOs

While many of the experimental searches for cold dark matter have been carried out by experimental physicists, astronomers also have been trying to solve the cold dark matter problem by measuring the abundance of entities called MACHOs—massive compact halo objects. MACHOs are incomparably more massive than either WIMPs or axions. Located in the spherical halo of the Milky Way, they might be planets, faint stars, brown dwarfs, white dwarfs, neutron stars, and/ or black holes.

Gravitational Lensing

In 1986, Princeton astrophysicist Bohdan Paczynski suggested that a MACHO in the galactic halo could be detected when it acts as a gravitational lens by bending light from a more distant astrophysical source as the light passes near the MACHO. According to the calculations of Paczynski and others, the exact details of a gravitational lensing event should enable astronomers to measure the mass of the object that acts as the gravitational lens. This would make gravitational lensing an enormously powerful new tool in the toolkit of astronomers. So what is it?

Were I to travel from New York City to Beijing, I could save money on fuel if I chose to travel the shortest physical distance between the two cities. The very shortest path through the three dimensions of space with which we are familiar would be a straight-line tunnel carved through the crust and mantle of the Earth. A tunnel to China, of course, does not exist, nor is it likely to ever exist. The shortest path that I might actually take would be by air, skimming just over the Earth's surface and crossing the North Pole. Since our flight path would be located only a few thousand feet above the Earth's surface, my "straight line" effectively would follow the curved surface of the Earth. In the language of mathematics, my flight path would be known as a *geodesic*, the absolute shortest possible path that connects two locations through the curved space through which we can actually travel.

One of the more mundane aspects of the physics of light is that light travels along the shortest possible path through space from point A to point B. Since space, like the surface of the Earth, is curved, that shortest possible path may not look to us like a straight line through flat space. One of the more dramatic predictions that emerged from Einstein's theory of general relativity, which is fundamentally a mathematical description of gravity, is that mass determines the shape of space. Specifically, in the presence of small amounts of mass space appears to be very flat, while in the presence of large amounts of mass space is measurably curved. Light cannot travel outside of space, so light will travel along the shortest available path through curved space. We could reasonably describe the light path as a straight line (i.e., the shortest path) through curved space. This is not, however, the language we normally use. Instead, we talk about mass bending the path of light, which makes the phenomenon sound, incorrectly, as if the light is traveling along curved lines through flat space. But no matter how we describe this phenomenon, the result is the same: light travels along the shortest possible path, the quickest route, through space that is sculpted by mass.

In 1919, Sir Arthur Eddington famously led an expedition to Principe, an island that straddles the equator near Gabon, just off the coast of western Africa, to conduct an experimental test of Einstein's theory of general relativity. Simultaneously, under instructions from Eddington, Andrew Crommelin led another British team to make identical measurements from Sobral, in northern Brazil. The idea was

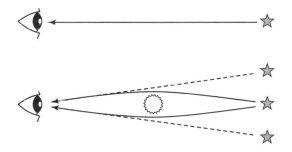

Figure 23.1. Illustration of gravitational lensing. Top, without gravitational lensing, light from a star travels in a straight line to the observer. Bottom, when a large mass lies in between the observer and the star, the observer cannot see the star at the star's true position because of the mass between them. However, light from the star is bent by the gravitational lens, so the observer sees multiple images of the star.

straightforward: on May 29, 1919, during a five-minute total eclipse of the Sun, Eddington and Crommelin and their teams would measure the positions of some stars in the Hyades star cluster, which during the moments of the eclipse would be located just beyond the dark edge of the solar disk. Since Einstein's theory predicts that the mass of the Sun curves space so that the space near the Sun acts like a lens to bend and focus light, these stars should be displaced outward by about one second of arc from the locations at which they should be found if Newton's, rather than Einstein's, theory of gravity were right. The Eddington and Crommelin expeditions were successful; in London on November 6, the Astronomer Royal, Sir Frank Dyson, reported the results to professional colleagues at a closed meeting of the Royal Society and the Royal Astronomical Society. On November 7, *The Times* of London, under a banner headline "Revolution in Science, Newtonian Ideas Overthrown," declared: "The scientific concept of the fabric of the Universe must be changed." Three days later, the *New York Times* offered the headline: "Einstein Theory Triumphs, Stars Not Where They Seemed or Were Calculated to Be." Einstein became world famous overnight.

In more recent decades, astronomers have observed numerous examples of massive objects bending light. This phenomenon is now called *gravitational lensing*; the massive object bending the light is called

Figure 23.2. The gravitational lensing effect known as Einstein's Cross, in which four images (top, bottom, left, right) of the same distant source are seen lensed by a nearby object (center). In this image, a galaxy only 400 million light years away is lensing a quasar located 8 billion light years away. Image courtesy of NASA, ESA, and STScI.

Figure 23.3. Gravitational lensing seen toward the galaxy cluster Abell 2218. The arcs are lensed images of galaxies located five to ten times farther than the lensing cluster. Image courtesy of W. Couch (University of New South Wales), R. Ellis (Cambridge University), and NASA.

the *gravitational lens* while we call the much more distant object whose light is bent and seen by us as one or multiple distorted objects the background source. Gravitational lenses that have been discovered in recent years include Earth-sized planets orbiting distant stars in the Milky Way as well as supermassive galaxy clusters in the relatively distant universe. The exact number and shapes of the distorted images of a background source seen in the vicinity of a gravitational lens depend on several factors, including the mass of the gravitational lens, the distance from Earth to the gravitational lens, the angular distance from the gravitational lens to the lensed image(s) of the background object, and the distance from the gravitational lens to the even more distant background. If the mass of the gravitational lens is large enough and the angle between the gravitational lens and the background object is small enough, we will see multiple images or a stretched out and curved image of the background object. This situation is known as "strong gravitational lensing." In other cases, the lens is not strong enough to generate multiple images or even a stretched and curved image of a background source but is strong enough to stretch the image of the lensed object. This situation is known as "weak gravitational lensing."

The MACHO Project

In 1992, the MACHO project team began a survey of twelve million stars in the Large Magellanic Cloud. Because of the direction of the LMC in the sky, astronomers on Earth have to look through the relatively nearby (10 kiloparsecs) halo of our own Milky Way in order to see stars in the relatively distant (55 kiloparsecs) LMC. Consequently, every now and again, a star in the Milky Way halo passes directly in front of a star in the LMC. During that brief interval of time, the halo star becomes a gravitational lens, a MACHO, and affects the paths of photons from the LMC star passing behind it. The MACHO project team watches to see if, over a period of a few days to weeks, any of the LMC stars brighten slowly and smoothly and then return to their normal brightnesses in exactly the same manner. Such a brightening event would signal gravitational lensing, meaning that the LMC star had passed directly behind a halo star. By monitoring the brightnesses of 12 million LMC stars, the MACHO project team obtained a statistically meaningful estimate of the number of unseen, faint halo stars in the Milky Way.

By 2000, the MACHO team had found a small handful of likely MACHOs and suggested, using statistical arguments, that MACHOs might make up as much as 20 percent of the dark matter mass of the Milky Way. A competitor collaboration known as EROS-2 monitored the brightnesses of more than thirty-three million stars in the Large and Small Magellanic Clouds and reported, in 2007, that the proportion of MACHOs in the galactic halo must be less than 8 percent of the dark matter mass of the Milky Way. Whether the correct figure is 20 percent or less than 8 percent, the population fraction of MACHOs in the galactic halo is much smaller than what is needed to provide the solution to the cold dark matter problem.

The Problem with MACHOs

Besides the fact that MACHOs appear to be far too scarce to solve the cold dark matter mystery, they suffer from another very significant problem: if the mass of the galaxy in the form of objects composed of normal matter (protons, neutrons, electrons) were ten times greater, which would be enough to solve the dark matter mystery, then the density of normal matter during the first few minutes of universal history must also have been ten times greater. As we will understand better in Chapters Twenty-four and Twenty-five, a higher density of normal matter in the early universe would mean more collisions of extremely energetic nuclear particles during the epoch when Big Bang nucleosynthesis was taking place. More proton on proton and proton on neutron collisions would mean that more deuterium (heavy hydrogen nuclei) should have been created more quickly; a higher abundance of deuterium in the first three minutes after the beginning would in turn mean that more helium should have been produced from deuterium on deuterium collisions, which ultimately would lead to a *lower* deuterium abundance; more helium would mean that more lithium and even some carbon, oxygen, and other heavier elements should have been produced from nuclear fusion events in the hot, young universe, before the density dropped below that necessary for nuclear fusion to continue. As a result, a solution (e.g., MACHOs) to the cold dark matter mystery that dramatically increases the number of protons and neutrons in the universe would be a solution that is otherwise inconsistent with the relative proportions of these elements that we have observed in the universe. The relative paucity of MACHOs is consistent with everything else we know about the universe, so the results of the MACHO and EROS-2 experiments confirm what we have learned from our other cosmological measurements; however, they do not lead us toward a better understanding of what the cold dark matter is.

To explain, consistently, the relative amounts of the elements astronomers have measured in the universe, the flat rotation curves for galaxies, and the high velocities of galaxies in clusters, cold dark matter must be more than dark and cold. It must be different. It must be matter that interacts via gravity with protons and neutrons in the universe but does not interact in any way with light; that is, cold dark matter must be made of stuff that does not respond to the electromagnetic force. This is why axions and sparticles are such appealing cold dark matter candidates. While experimental searches on Earth for axions and sparticles have thus far come up empty, powerful evidence has emerged from some astrophysical measurements—specifically gravitational lensing experiments using X-ray and visible light telescopes in space—that clearly points to the existence of an enormous amount of cold dark matter. Since this cold dark matter cannot be made of normal matter, it must be made of exotic dark matter, perhaps axions or sparticles, or perhaps some other kind of matter about which we do not even know enough to speculate as to what it might be.

The Bullet Cluster

One very special place where the gravitational lensing technique has been applied and the question of the existence of cold dark matter has been put to the test is the Bullet Cluster. The Bullet Cluster actually consists of two clusters of galaxies that collided about 150 million years ago at a relative speed of 4,500 kilometers per second, which is about 1.5 percent of the speed of light. In thinking about the collision, we need to consider what *should* have happened to the main components of the cluster during and as a consequence of the collision, depending on whether cold dark matter does or does not exist. One component is comprised of those constituents of galaxies that generate substantial amounts of visible, infrared, and radio light, namely, the stars and clouds of gas within every galaxy in each of the two galaxy clusters; they are made of normal matter. The second component is the million-degree intergalactic gas, most of which is hydrogen, that can be seen in X-rays and which is also made of normal matter. The third component is the dark matter that, if our interpretation of flat rotation curves for spiral galaxies is correct, comprises as much as 80 percent of the internal mass of every galaxy, though we do not know whether that dark matter is made of axions, WIMPs, MACHOs, or some completely unknown material. We can say with complete confidence that the first two components exist. But the third component? It must exist if nothing other than the existence of a large amount of unseen mass can explain the flat rotation curves of galaxies; however, some physicists have argued that our theory of gravity is incomplete and that a small change

Figure 23.4. This composite image shows the galaxy cluster 1E 0657-56, also known as the Bullet Cluster. This cluster was formed by the collision of two large clusters of galaxies, the most energetic event known to have taken place in the universe since the Big Bang. The two large white-gray patches closest to the center show the locations from which X-rays are being emitted from hot gas. The outer two white-gray areas, which overlie the two locations of the visible galaxies, show the locations of the mass responsible for gravitational lensing of more distant galaxies. X-ray image courtesy of NASA/CXC/CfA/M.Markevitch et al. Optical image courtesy of NASA/STScI; Magellan/U.Arizona/D.Clowe et al. Lensing Map courtesy of NASA/STScI; ESO WFI; Magellan/U.Arizona/D.Clowe et al.

to Newton's laws (called Modified Newtonian Dynamics, or MOND) could explain flat rotation curves without resort to dark matter. Maybe, say MOND proponents, astrophysics tell us nothing about whether axions and sparticles are real. Observations of the Bullet Cluster have provided the definitive rebuttal to MOND.

Remember that the physical sizes of the visible parts of galaxies are a few tens of thousands of light-years across (the visible part of the Milky Way has a diameter of about 100,000 light-years), which are very small distances in comparison to the millions-of-light-year distances between galaxies within a cluster (the distance from the Milky Way to

Andromeda is about 2.5 million light-years). Similarly, the wingspan of a single goose in a flying flock of geese is typically much smaller than the separation between the wing tips of any two geese.

Now imagine that two flocks of geese are flying in opposite directions over the Willis Tower (formerly the Sears Tower) in downtown Chicago. In such a situation, no two geese are likely to collide; the two flocks pass right through each other with nary a bump or an angry honk. Next imagine that one of our flocks of geese is migrating northward toward Canada. Because of a high-pressure system in the Gulf of Mexico, hot moist air from the Gulf is being pushed northward with the flock of geese at the same speed and in the same direction as the geese are flying. In effect, the geese are embedded within a large bubble of hot, moist air, and both the geese and the bubble of hot air are traveling northward together. Meanwhile, our other flock is southbound, headed to Florida, but because of a high-pressure system in the Arctic that is pushing cold air to the south, these geese are embedded within a large bubble of Arctic air and are flying southward at the same speed and in the same direction at which the cold front is moving to the south. In effect, each "flock" consists of two kinds of matter: animate matter, consisting of a few hundred geese, and a bubble of inanimate matter, consisting of countless, invisible air molecules.

Again, let's imagine that the two "flocks" meet over the Willis Tower. Again, the geese fly right past each other, but when the bubbles of hot and cold air pass over the tower at the same time as the geese, the air bubbles collide violently. They are unable to pass though each other so, after smashing into each other, they come to rest. Now a weather front is suspended over Chicago. The crash of thunder and the flash of lighting fill the Chicago sky. Despite the collision of the air bubbles, which has stopped the hot and cold air parcels over Chicago, the geese keep on flying, one flock up to Canada and the other down toward Florida. In a few hours, one flock of geese is winging its way further north over Wisconsin, the other continues its southward migration over Indianapolis, while rain falls on Chicago.

The visible galaxies in the two clusters, like the geese in our two flocks, sail right past each other, with each galaxy barely noticing the presence of any other. The galaxies do interact through gravity; they pull on each other a little bit. Some might even be distorted or stretched by these gravitational interactions, but the galaxies are so far apart and moving past each other so fast, that the gravitational tugs are small and thus have almost no effects, either immediate or long-term. But remember that each cluster of galaxies is filled with a cloud of million-degree gas that fills up the entire volume of space between the galaxies in each cluster. When the clusters try to pass through each other, the clouds of million-degree gas smash into each other, like a hot front meeting a cold front in the atmosphere of the Earth. As a result of the head-on collision, the two clouds are stopped in their tracks while the galaxies themselves keep on moving in the directions in which they were originally traveling. Today, 150 million years after the galaxy clusters first collided, X-ray-bright gas clouds remain marking the site of the collision while the visible galaxies have continued on their separate paths.

This is where the Bullet Cluster gets interesting. Remember that in a typical galaxy cluster, the total mass of gas evident as hot, intergalactic, X-ray-emitting plasma is ten times greater than the amount of ordinary mass generating the visible, infrared, and radio light from the stars and cold gas clouds in the galaxies. The Bullet Cluster is no exception: the X-ray-emitting plasma, which has now been removed from both clusters of galaxies, contains about ten times more mass than the total mass of the stars and gas clouds in the galaxies. One hundred and fifty million years post-collision, the Bullet Cluster consists of two main regions. The central region, revealed in the hot gas seen in Chandra X-ray Telescope images, contains about 90 percent of the ordinary matter; the periphery shows two clusters of galaxies revealed in Hubble Space Telescope images, one on either side of the hot gas region, which together contain about 10 percent of the ordinary matter.

If the Bullet Cluster were to act as a gravitational lens of more distant galaxies, the volume of space containing the greatest amount of mass (the center or the periphery?) should act as the dominant gravitational lens. Given the observed distribution of normal matter, and if all the mass is ordinary matter in the hot gas and the two clusters of galaxies, and since the X-ray-emitting gas contains ten times as much mass as the visible galaxy clusters, the hot gas cloud should serve as the center of the lensing environment. Amazingly, observations reveal just the opposite.

In 2004, Douglas Clowe, then at the Steward Observatory in Arizona, led a team that used the Hubble Space Telescope to study the weak gravitational lensing effects of the Bullet Cluster. They found that the region *associated with the visible galaxies* is responsible for the gravitational lensing. In fact, about 80 percent of the total mass of the Bullet Cluster is associated with the region of space where the visible clusters of galaxies are located; yet, the normal matter producing the light we see when we observe the galaxy clusters contains ten times less mass than the X-ray-emitting plasma. The only viable explanation is that an enormous amount of dark matter still surrounds and/or is inside of the galaxies within the clusters. Furthermore, whatever this dark matter is, it interacts gravitationally (producing gravitational lensing effects) but is otherwise effectively collisionless: when the galaxy clusters collided, the dark matter components passed right through each other, as did the visible galaxies. MOND is wrong. Collisionless cold dark matter is real.

These observations of the Bullet Cluster provide us with new information about the three mass components of a galaxy cluster. The first component, which provides at most 1 to 2 percent of the mass of the cluster, is the normal matter that makes up the stars and clouds of gas that light up the galaxies. The second, which accounts for ten times more mass than the first, i.e., perhaps 10 to 20 percent of the total cluster mass, is also composed of normal matter in the form of hot intracluster gas that emits copious amounts of X-rays. The third, which accounts for 80 to 90 percent of all the mass in the cluster, is the cold dark matter that surrounds and permeates each visible galaxy. And what makes up all the cold dark matter? We still do not know.

In 2008, a team lead by Maruša Bradač, of UC Santa Barbara, studied another massive merging galaxy cluster and found another "Bullet Cluster" that yielded the same results: the stars (the visible galaxies) comprise about 1 percent of the mass of the colliding galaxy cluster system, intergalactic gas makes up about 9 percent of the total, while collisionless dark matter makes up the remaining 90 percent.

The two "Bullet Clusters" provide extremely powerful visual evidence for the presence of all three of these distinct components including, most importantly, enormous amounts of cold dark matter that does not collide with or otherwise interact with the normal matter. In fact, the collisionless cold dark matter is by far the dominant component, by mass, of these two merging or colliding galaxy clusters. We just don't know what or how exotic the cold dark matter is.

More Evidence of Dark Matter: The Missing Dwarf Galaxies

Most astrophysical theorists believe the formation of galaxies was triggered by the enormous gravitational influence of dark matter particles. The majority would go so far as to conclude that galaxies are gravitationally bound clumps of invisible dark matter that also contain much smaller amounts of normal matter, some of which we can see. In fact, according to most theorists, the universe does not contain enough normal matter for galaxy formation to have ever happened without the initial presence of large amounts of dark matter.

In the ancient past, dark matter began to clump; the clumping of dark matter triggered the clumping of normal matter to form protogalaxies, the seeds of modern galaxies. Once protogalaxies, containing both dark and normal matter, formed, the normal matter fell to the center of the dark matter clumps, leaving the dark matter in halos around central cores of normal matter. Once star formation began and nuclear fusion inside stars provided a source of light, the normal matter cores became the luminous galaxies we see today.

Some theorists also predict that large spiral galaxies like the Milky Way should be surrounded by dozens of very small galaxies, many of which would contain very little normal matter and therefore very few stars. In effect, these clouds of small, dark matter-dominated companion galaxies are the residue of the original phase of galaxy formation. Until recently, though, astronomers had found and identified almost no such dwarf galaxies. That situation has changed as the number of known dwarf satellite galaxies of the Milky Way has doubled with discoveries made in the first decade of the twenty-first century. These newly discovered dwarf galaxies are especially interesting because they appear to contain relatively few stars and consequently are much fainter than any previously known dwarf galaxies. But do these ultra-faint dwarf galaxies contain few stars because they have very little mass? Or is it because, as theorists have predicted, these galaxies are made up almost entirely of cold dark matter, which creates an ineffective environment for the process of star formation?

When astronomers study galaxies, one measurement they try to make is the ratio of the amount of mass in the galaxy to the light output of the galaxy. The Sun has a mass-to-light ratio of one (measured in units that make sense for stars and to astronomers—solar masses divided by solar luminosities—rather than in units of grams divided by watts). If a galaxy were made of 1,000 stars identical to the Sun in mass (1,000 solar masses, total) and each star were identical to the Sun in brightness (1,000 solar luminosities, total), the mass-to-light ratio would be one, since the numerical value of 1,000 solar masses divided by 1,000 solar luminosities is equal to one; however, if a galaxy has 1,000 stars but each star has a mass of only half that of the Sun (500 solar masses, total), and if each of these stars is only one-tenth as bright as the Sun (100 solar luminosities, total), the mass-to-light ratio for this galaxy would be five. If a galaxy has a large mass-to-light ratio, then much of the mass in the galaxy is ineffective or plays no direct role in producing light.

In the inner part of the Milky Way, where stars are dominant and dark matter plays a very limited role, the mass-to-light ratio is about ten, which means that the galaxy needs the mass of about ten stars like the Sun to produce as much light as a single Sun. Some of this mass is in clouds of gas in between the stars, some of it is in stars, and some of it is cold dark matter. For the Milky Way as a whole, the mass-to-light ratio is in the range of 30 to 100, with the total mass dominated by cold dark matter. In clusters of galaxies like the Coma Cluster, in which Zwicky detected the first definitive evidence for dark matter, the mass-to-light ratio can be as large as a few hundred. In 2007, astronomers using data from the Sloan Digital Sky Survey discovered galaxies with even larger mass-to-light ratios. These galaxies, called *ultra-faint galaxies*, have been found as satellite galaxies of the Milky Way. Their mass-to-light ratios range from a few hundred up to nearly two thousand, indicating that they are overwhelmingly dominated by matter that has not formed stars. One plausible explanation for this phenomenon, is that this matter is cold dark matter that is unable to form stars.

The Case for Dark Matter

A century ago, astronomers thought they knew what the term "dark matter" meant. It was matter that produced so little light that it was hard to see. But as astronomers' tools improved, much of that not-too-luminous matter, whether found as hot gas in disks around black holes or as white dwarfs or distant planets or faint stars or hot intergalactic gas, became visible. Now, dark matter is defined a bit differently.

In modern terminology, the definition has expanded. It still includes normal matter (composed primarily of protons, neutrons, and electrons, which astronomers call, collectively, call *baryonic matter*) that exists in forms that make it hard or impossible to see. Black holes, neutron stars, white dwarfs, and hot intergalactic gas are examples of this category of dark matter. But the definition now also includes matter that is "exotic" in that it does not behave in a "normal" manner. It affects other objects through the force of gravity but does not interact with other objects in any other way (or interacts so weakly that the non-gravitational interactions are negligible compared to the gravitational interactions). Such objects would be composed of non-baryonic dark matter (particles not made from protons, neutrons, and electrons) and could be in the form of axions, WIMPS, or other particles or objects we currently lack the imagination and knowledge of to discover.

The evidence for the existence of non-baryonic dark matter with these qualities is now overwhelming. It includes the too-fast motions of galaxies in clusters of galaxies, the too-fast orbital velocities of stars and gas clouds in spiral galaxies, the gravitational lensing seen toward colliding galaxy clusters, and the extremely high mass-to-light ratios of the ultra-faint galaxies that orbit the Milky Way. It also includes the fact that in the early history of the universe galaxies could not have formed at all if the gravitational influence of non-baryonic dark matter had not triggered the process, as well as the requirement, deduced from the observed relative amounts of the elements and the physics of nucleosynthesis in the hot, young universe, that almost none of this mass can be normal matter.

In all of the environments in which dark matter must be present, the dominant form of matter shows no evidence that it produces any light or is capable of colliding with normal matter; it reveals its presence through the ways it sculpts space via the gravitational force. But what is it? Most of it must be exotic, and while we have some candidates in WIMPs and sparticles, in these early years of the twenty-first century we still cannot answer that question. We do not know what kinds of particles comprise exotic dark matter. What we do know is that the existence of this dark matter is crucial for understanding how to determine an age of the universe from the cosmic microwave background.

· · · CHAPTER 24 · · ·

Hot Stuff

"Well boys, we've been scooped."

—Robert Dicke, 1965, as quoted in R. B. Partridge, *The Cosmic Microwave Background Radiation* (1995)

Just before Hubble found evidence that the universe was expanding, the Belgian priest and physicist Georges Lemaître and, completely independently, the Russian physicist Alexander Friedmann suggested that a universe governed by Einstein's law of gravity, general relativity, had to be either expanding or contracting. Friedmann's contribution was to show that Einstein's static spherical universe was dynamically unstable; any small disturbance would cause it to expand or contract. While his work was almost purely mathematical, Lemaître's research was directly related to physical cosmology. He noted that the evidence shows that the universe we live in is not contracting and therefore must be expanding—"the receding velocities of extragalactic nebulae are a cosmical effect of the expansion of the universe"—and that, as is self-evident once one thinks about the situation seriously, an expanding universe must have been smaller in the past. In the most remote past, the universe must have been unimaginably small, and the instant that marks the beginning of time must have been the instant when it was at its smallest. He wrote, "We could conceive the beginning of the universe in the form of a unique atom, the atomic weight of which is the total mass of the universe." This work of Lemaître, published in the obscure Belgian journal *Annales de la Société scientifique de Bruxelles* in 1927, marked a shift in intellectual history. The idea that the universe

Figure 24.1. Left, Belgian priest and physicist Georges Lemaître. Middle, British astrophysicist Arthur Eddington (seated, left), with physicists Henrik Lorentz (seated, right) and Albert Einstein, Paul Ehrenfest, and Willem de Sitter (left to right, standing). Right, Russian-American astrophysicist George Gamow. Photos courtesy AIP Emilio Segrè Visual Archives (center). and The George Washington University Library (right).

had a beginning was no longer confined to theological speculation. Once and for all it had entered the realm of cosmology.

In 1930 Lemaître sent a copy of his paper to Sir Arthur Eddington, who announced Lemaître's results to the world via a letter to the journal *Nature* (June 1930) and arranged for an English language translation of Lemaître's paper to be published in *Monthly Notices of the Royal Astronomical Society* in March of the following year. Eddington himself rejected the idea that the universe could ever have been so unimaginably small; he wrote in his *Nature* letter that "the notion of a beginning of the present order of nature is repugnant to me." Later that year, in an October letter to the journal, he explained that he accepted the expansion of the universe but believed that the universe had never been small and had no beginning in time: "Its original radius was 1,070 million light-years, before it began to expand." Lemaître responded that "the beginning of the world happened a little before the beginning of space and time. I think that such a beginning of the world is far enough from the present order of Nature to be not at all repugnant." Despite Eddington's reservations (which appear to have been inspired by notions that lie outside of the domain of science), the generalnotion that the physical universe did have a beginning at a specific moment in time took hold, and it remains fundamental to modern cosmology.

Over the next decade and a half, quantum mechanics came to dominate the world of theoretical physics, and physicists working on the Manhattan Project deciphered the basic physics of both the atom and of the process by which stars power themselves. The physicist George Gamow, who was born in Russia in 1904 and naturalized as a citizen of the United States in 1940, did not work on the Manhattan Project; he was a professor at George Washington University in Washington, DC, doing theoretical research on how elements of the periodic table formed in nuclear fusion reactions inside stars. In 1942, he proposed the idea that the origin of the elements, from lightweight hydrogen to heavyweight uranium, might be understood as part of the explosive fragmentation process that must have taken place in a young universe of extreme heat and density. "No composite nuclei could have existed under these conditions, and the state of matter must be visualized as a hot gas formed entirely of nuclear particles, that is protons, neutrons, and electrons," he wrote in his 1952 book *The Creation of the Universe*.

By 1946 Gamow realized that fusion in a young, hot, dense universe, even one as hot as 10 billion degrees, would not yield heavy elements. Since protons have like charges, they cannot be bound together into nuclei without the help of neutrons, and since free neutrons are unstable and spontaneously decay into protons and electrons in only a few minutes, the ingredients for building heavy elements in a hot, small universe would have quickly disappeared. In addition, as he worked through the details of the physical processes at work in the young universe, he found that both the temperature and density of an expanding universe would decrease. Since nuclear fusion requires enormous temperatures and densities to work, it would quickly cease operation in an expanding, cooling universe. Instead of all of the elements, only the lightest—hydrogen, helium, and lithium—would have been forged in the nuclear fusion furnace that was active during the earliest epochs in the history of the universe; as a result, the heavier elements must have formed later, in nuclear reactions inside stars.

The Cosmological Redshift

Recall that in 1915, Willem de Sitter discovered, in solving Einstein's equations, that the stretching of the space of the universe would also stretch the photons. Think of a photon as a sleeping snake that is not moving through space and that has a head and a tail separated by a span of space equal to the wavelength of the photon. Since space is being stretched, and since the head and tail of the snake are fixed at specific locations in space, the distance between the head and tail of the snake must increase. For our photon, this is equivalent to stretching the wavelength. Matter itself is not stretched by the stretching of space because the electromagnetic and gravitational attraction of one piece of mass for another on a small spatial scale in the universe (a penny, the earth, the solar system, the Milky Way galaxy) is overwhelmingly more powerful than the stretching of space; the myriad particles that make up objects hold themselves together and preserve their physical sizes while the fabric of space gets stretched. However, photons, being massless particles, do get stretched.

In fact, the wavelengths of photons expand at exactly the same rate as the expansion of the universe as a whole. If the universe doubles in size, the wavelengths of photons also double in size. While neither the current nor the past size of the universe is directly measurable, the ratio of its current to its past size can be measured and is of interest to us since this ratio tell us the factor by which the universe has changed in size since the moment long ago when an astrophysical object we see emitted the light we collect with our telescopes today.

If a hydrogen atom emitted a photon of wavelength 656.2 nanometers when the universe was one percent of its current size, by now that photon would have been stretched to 100 times its original size. It would have a wavelength of 65,620 nanometers. The photon, originally emitted at a red wavelength detectable by human eyes is now redshifted into the far-infrared where it is invisible to our eyes. This redshift, which is caused by the expansion of the universe rather than by the physical motion of an object away from us through space, is called a *cosmological redshift*. These are the redshifts originally observed by Vesto Slipher in 1913 and by Edwin Hubble and Milton Humason in the late 1920s.

If for a very distant object we determine the value of that object's cosmological redshift, we immediately know how much bigger the universe is now than it was when the just-observed photons from that distant object were emitted. For example, if the cosmological redshift is 0.3, the universe is 30 percent bigger now than when these photons

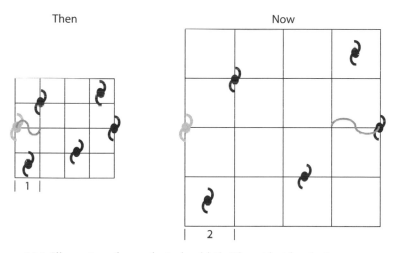

Figure 24.2. Illustration of cosmological redshift: The grid with galaxies represents the universe. On the left, the universe is shown as it was in the past, at a moment when a photon is seen leaving the left-most galaxy. On the right, the universe is seen at the present time, when the photon reaches the right-most galaxy. From "then" until "now," the entire universe has doubled in size, so that the distances between galaxies are twice as large and the photon itself is twice as large. The galaxies, however, have not moved from their original locations in the universe and have not changed in size.

were emitted from this object located in a part of the universe very far distant from us.

The numerical value of the measured cosmological redshift also yields the velocity by which two distant objects are receding from each other due to the expansion of the universe. In 1929 and 1931, Hubble was measuring cosmological redshift values of about 0.003 to 0.03, which correspond to cosmological redshift velocities in the range of about 1,000 to 10,000 kilometers per second.

The Universe Cools Off

When the universe was extremely young, small, and dense, it was opaque because the photons within it were continually bouncing around, principally being scattered by free electrons. No photon could travel far before such an event would occur. The temperature of such a universe, at any moment, would be determined by the distribution of photons across all wavelengths as they bounced around, in the same way that the temperature of the air in a room or the temperature of gas in the atmosphere of a star is determined by the distribution of velocities of the particles in those environments. If we measured this distribution—how many photons exist at each wavelength—we would be measuring a blackbody distribution with a temperature that can be directly calculated from Wien's law. Remember from the definition of Wien's law that the wavelength at which a blackbody emits the maximum amount of light yields the temperature of the object (Chapter Seven).

If all the photons in the universe are stretched enough, then X-ray photons become extreme-UV photons, blue photons become red ones, and near-infrared photons are shifted into the far infrared. Consequently, as a result of the expansion of the universe and the stretching of all the photons contained within it, the entire blackbody distribution is shifted toward longer wavelengths, and the peak of the distribution will be shifted to a longer wavelength. Since a blackbody that peaks at a long wavelength has a cooler temperature than a blackbody

Figure 24.3. Left, a composite picture, dated 1949, showing Robert Herman (left), Ralph Alpher (right), and George Gamow (center). Gamow is seen as the genie coming out of the bottle of "Ylem," the initial cosmic mixture of protons, neutrons, and electrons from which, Gamow hypothesized, the elements formed. Right, Robert Dicke. Image courtesy of Princeton University Library (right).

that peaks at a short wavelength, Wien's law tells us, *the result of stretching all of these photons is that the universe cools.*

Gamow had this insight in 1946 and subsequently put his graduate student, Ralph Alpher, to work on answering the question of how nucleosynthesis would have operated in a very hot, young universe as it expanded and cooled off. In 1948, Gamow and Alpher published "The Origin of Chemical Elements" in the journal *Physical Review*, in which they demonstrated that the physical reactions in the first moments of the universe, when the universe was still hot and dense enough to permit fusion reactions to happen, would naturally produce a universe made almost entirely of hydrogen and helium.

Initially the universe, as described by Gamow and Alpher, was pure energy. Energy, however, as shown by Einstein, is equivalent to mass multiplied by the square of the speed of light, $E = mc^2$. As a consequence of this equivalence of energy and mass, energy in the form of light (photons) can turn into energy in the form of mass (particles). The reverse process, in which mass turns into light, also can occur. Since shorter wavelength photons (for example, X-rays) have more energy than longer wavelength photons (for example, radio waves), a shorter wavelength photon (higher energy = higher equivalent mass) can turn into a more massive particle than can a longer wavelength (lower energy = lower equivalent mass) photon. Consequently, as the universe cooled and the photons were stretched, the particles that could spontaneously appear (when a photon changed into a particle) and disappear (when a particle changed into a photon) transitioned from higher to lower-mass particles. Eventually, the photons became so stretched, and therefore the universe so cool, that they could no longer turn into particles. They lacked enough energy to change into even the lowest-mass particles.

As long as the temperature of this expanding, cooling universe remained above one thousand billion degrees, or 10^{13} K, some of the photons spontaneously could turn into protons and neutrons and electrons, and an equal number of protons and neutrons and electrons spontaneously could turn back into photons. Equally important, because of a very subtle bias in the laws of physics, the universe created slightly more protons and electrons than anti-protons and anti-electrons (the anti-particles are identical to the particles in every way except their electric charge). After all the available anti-protons combined with protons and by mutual annihilation turned from matter to energy (i.e., became photons), only protons and neutrons remained from the sea of massive particles. At a certain point, about one millionth of a second after the beginning, when the photons were stretched enough that the temperature of the universe plunged below 10^{13} K, the stretched photons no longer had enough energy to turn back into the massive particles, and vice versa, so the number of protons and neutrons in the universe froze; later, about five seconds after the beginning and as the temperature dropped below about 6 billion degrees, the photons could no longer turn into less massive particles like electrons and anti-electrons and those particles could no longer turn back into photons. At this moment, five seconds after the beginning, after all the available anti-electrons had annihilated by combining with electrons and turning into photons, all the protons, neutrons, and electrons—the fundamental building blocks of atoms—that would ever exist had been created. Not all of them would survive.

Nucleosynthesis in the Hot, Young, Dense Universe

Neutrons are not stable on their own but are stable if they are bound into a nucleus with other protons and neutrons. Free neutrons, these being neutrons that are not bound into a nucleus, have a half-life of only 887.5 seconds (a little less than fifteen minutes), and so in the young universe free neutrons began to fall apart into protons and electrons. On the other hand, any neutron that collided with and paired with a proton to form a heavy hydrogen nucleus (deuterium) survived. In the first few minutes, after the formation of some deuterium nuclei, some protons and deuterium nuclei collided with sufficient energy to combine into helium nuclei (two protons plus one or two neutrons) and then some protons and helium nuclei collided and combined to form lithium nuclei (three protons plus three or four neutrons). But as the universe expanded further, the particles lost more and more energy. In addition, particles with two (helium) or three (lithium) protons require much higher speeds (i.e., hotter temperatures) in order to collide with each other with enough energy to trigger fusion reactions. Soon the nuclei were moving too slowly to react at all. The formation of light elements came to a screeching halt; no elements heavier than lithium formed at all. After only a few hours, all the free neutrons were gone. By that time, the hot, expanding universe was filled with photons, electrons, protons, and nuclei of deuterium, helium, and lithium, but no neutrons that were not locked into nuclei with protons. The universe was now awash with interacting particles that could no longer change back and forth into photons and with photons that could no longer change back into particles. The universe, of course, kept on expanding and cooling.

In this universe of particles, a cycle of events began. First, those particles that were positively-charged (protons; deuterium, helium, and lithium nuclei) encountered those that were negatively-charged (elec-

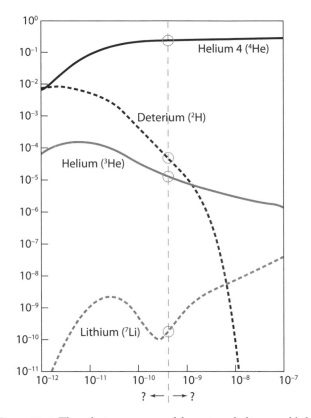

Figure 24.4. The relative amounts of deuterium, helium, and lithium depend on the overall density of normal matter during the first few minutes of the history of the universe, when Big Bang nucleosynthesis took place. For example, if the density were low (left), very little lithium would be produced; if the density were high (right), most of the deuterium and ^{3}He would fuse into lithium and ^{4}He. The observed relative amounts (the circles on each curve) of deuterium, helium, and lithium are all consistent with the same density for ordinary matter. Image courtesy of NASA.

trons) and joined to become charge-neutral atoms. As a result of each combination reaction, some energy was given off as a photon. Next, the photons emitted in these combination reactions quickly found themselves colliding with free electrons (electrons not bound with protons into atoms). Just as would happen when a gust of cold air comes through an open door into a room full of warm air and the collisions of

TABLE 24.1.
The Early Universe

Age of the Universe	Temperature	Particles	What is Happening
1.5 millionths of a second	1,000 billion K	• number of protons and neutrons frozen	• below this temperature, these particles are too massive and can no longer turn back into photons
5 seconds	6 billion K	• the number of electrons is frozen • photons cannot travel far before interacting with electrons	• below this temperature, these low mass particles can no longer turn back into photons • the universe is opaque
3 minutes	1 billion K	• protons are stable • neutrons decay • there are seven times as many protons as neutrons • all remaining neutrons are stable in deuterium or helium nuclei • 75% of the total mass of all protons and neutrons are in hydrogen, 25% in helium	• protons and neutrons collide to form deuterium nuclei • deuterium nuclei and protons collide to form helium nuclei • nuclear fusion ends
380,000 years	3,000 K	• neutral hydrogen atoms • neutral helium atoms	• electrons combine with hydrogen and helium nuclei (recombination) • the universe becomes transparent • CMB is released

air molecules bring all the molecules to the same temperature, these photon-electron collisions redistributed the energies among the photons and electrons so that the sea of photons took on the same temperature as the sea of particles. Now, the photons made up a gas that permeated the universe and had a characteristic temperature. Finally, after being randomly scattered enough times by the electrons, these photons found themselves on paths that caused them to collide with neutral atoms. In the next instant, an electron that had just recently released a photon in order to combine with a proton into a neutral atom, absorbed a photon from the photon gas and found itself a free electron again. This cycle continued, with the charged particles finding their charged counterparts, then combining and releasing a photon, with that photon merging into the blackbody radiation (the photon gas, characterized by a specific temperature), and with another blackbody photon being absorbed by an electron in a neutral atom, causing that atom to dissociate into an electron and a positively-charged nucleus. On and on, this cycle continued. For as long as the blackbody radiation had enough energy to rev up electrons in atoms and turn them into free particles again, and for as long as enough free electrons existed in the cosmos that the photons could be guaranteed of continually scattering until they encountered a neutral atom, this process continued.

The Cosmic Microwave Background (CMB) Predicted and Discovered

This cyclical transformation of neutral atoms into free ions and photons and then back again into neutral atoms continued for about 380,000 years; however, with each cycle more time passed, and as time passed the universe grew in size. As the universe grew in size, the ambient blackbody photons grew in wavelength as the photons experienced greater and greater cosmological redshifts and consequently gradually had less and less energy. Since the sea of photons and the particles of matter were thermally coupled through their collisions, the matter cooled down, also. As the temperature of the matter cooled below 3,000 K, the cycle was broken. The electrons and positively-charged nuclei became bound into stable atoms, because the blackbody photons were stretched so far that they lacked sufficient energies to re-ionize the hydrogen atoms. Now, almost all of the electrons remained bound into neutral atoms, leaving 10,000 times fewer free electrons than before available to scatter photons. With so few electrons to scat-

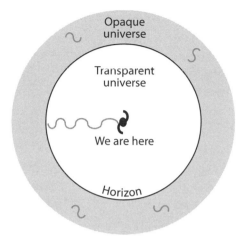

Figure 24.5. The cosmic microwave background consists of photons that were emitted when the universe was about 380,000 years old. In this cartoon, we look outwards from our location in the universe, and when we look outwards we look back in time. Photons emitted about 13.7 billion years ago have been redshifted during their journey toward Earth. Any photons created and emitted in the universe before 13.7 billion years ago cannot penetrate the horizon, as they were absorbed by other particles before they could travel far. On the other hand, at the moment the CMB photons were emitted, the universe was transparent, so those photons have traveled without being stopped or absorbed across 13.7 billion years of time and space to reach our telescopes.

ter them, the possibility of a photon encountering a neutral hydrogen atom dropped effectively to zero. A photon was more likely to travel the entire length or breadth of the universe than to see an atom in its path. This transition period, when the temperature of the universe was about 3,000 K and the photons were suddenly released, free to travel unimpeded across the universe, was the epoch when the universe became transparent to light. All the photons in the universe released during this transition period from every location in the universe, taken together, look like light released from a blackbody with a temperature of 3,000 K. They are still traveling. We call these photons the *cosmic background radiation*.

In 1948, Alpher and Robert Herman, Alpher's colleague at Johns

Hopkins, made a truly remarkable prediction. They predicted that the cosmic background radiation photons, the remnant afterglow signature of the origin of the universe, must still be present in the universe and should have a characteristic temperature. They calculated that with the expansion of the universe, after a few billion years these photons released from a 3,000 K blackbody should have a characteristic temperature of about 5 degrees above absolute zero (5 K). Over the next eight years, Alpher, Herman, and Gamow, would refine Alpher's and Herman's original calculations, using new observational data (not all of it good), and predict temperatures of 28 K (Alpher and Herman in 1949), then 3 K (Gamow in 1950), 7 K (Gamow in 1953), and 6 K (Gamow in 1956). Alpher left the field of astrophysics in 1955 (he went to work for General Electric); Herman followed in 1956 (he took a position with General Motors), and their prescient work was largely ignored for most of two decades. In his 1995 book *The Very First Light*, John Mather, who shared the 2006 Nobel Prize in Physics for his work on the cosmic background radiation, suggested a number of reasons why their work did not arouse greater interest. Foremost among them is the fact that the idea championed by Gamow of a primordial explosion that expanded and cooled was not yet serious, mainstream astrophysics in the early 1950s. It failed to explain the existence of the elements other than hydrogen and helium and the age of the universe according to the observed Hubble expansion law (2 billion years) was significantly less than the age of the oldest then-known Earth rocks.

In 1964, Robert Dicke and Jim Peebles of Princeton University, without knowing about the work of Alpher, Herman, and Gamow, rediscovered their predecessors' ideas on the physics of the early universe. According to the calculations done by Peebles, the present temperature of the cosmic background radiation should be about 10 K. At 10 K, the peak wavelength of the cosmic background radiation photons should be at 0.029 centimeters. Based on this prediction, most of the photons from the cosmic background radiation would be millimeter wavelength photons, but a significant fraction would be the centimeter wavelength photons that radio astronomers called microwaves. Since the first measurements of the cosmic background radiation made by radio astronomers were of microwaves, the cosmic background radiation is referred to as the *cosmic microwave background*, or CMB. Dicke and Peebles collaborated with their colleagues Peter Roll and David Wilkinson to build a radio telescope and radio wave (i.e., microwave) detector, which were to be set on the roof of a building on the Princeton campus where they would be used to try to detect the photons from the CMB.

Meanwhile, just 50 kilometers down the road from the Princeton campus, Arno Penzias and Robert Wilson, two researchers at the Bell Telephone Company research laboratory in Holmdel, New Jersey, were making measurements with a device originally designed to detect radio waves with a wavelength of 11 centimeters. This device, known as a radiometer, was originally built in 1961 by Bell Labs physicist Ed Ohm for testing satellite-to-ground communications with Bell Telephone Laboratory's Telstar satellite system, for which two satellites were launched, the first in 1962, the second in 1963. The first step toward satellite communications had been taken when NASA launched the first Echo satellite in 1959. Echo was a passive satellite that reflected telecommunications signals back to receivers on the ground. In preparation for working with the Telstar satellites, Ohm measured the reflected radiation from Echo and found too much signal in his measurements; he detected the expected telecommunications broadcast signal plus some extra energy. That extra energy, manifested as noise that Ohm and his co-workers needed to get rid of in order to generate a noise-free communications signal, was equivalent to a source of radio wavelength photons with a temperature estimated at ~3.3 K. Two Russian astrophysicists, Andrei Doroshkevich and Igor Novikov, recognized the connection between Ohm's measurements and Gamow's predictions, but most scientists in the West were unaware of either Ohm's work or the suggestion made by the Russian team.

By 1963 Bell Labs had decided not to continue using Ohm's radiometer for the Telstar satellite telecommunication project. Penzias and Wilson took the instrument, then unused, and turned it into a radio telescope for measurements at a wavelength of 7.4 centimeters, calibrated it, and began to do radio astronomy, first studying the halo of the Milky Way and then the supernova remnant known as Cassiopeia A. Their measurements were extremely reliable compared to those of

Ohm, but they also could not get rid of the problematic excess radiation. It did not seem to come from their equipment; it did not come from the Moon or from any other specific astrophysical sources; it did not come from nearby New York City or from the slightly more distant Van Allen radiation belts; it did not come from pigeon droppings inside the radiometer. And it made no difference what direction they pointed their radio telescope, the noise was simply always there. Penzias and Wilson estimated that this all-sky noise source had a temperature of 3.5 ± 1 K.

In early 1965, Penzias mentioned the problem in a phone conversation with astrophysicist Bernard Burke. Burke, who had recently read a draft of a paper written by Peebles in which Peebles made the prediction that relic photons from the early universe might be detectable as thermal radiation from all directions in the sky, suggested that Penzias contact Peebles. Penzias phoned Princeton in March and spoke to Dicke while Dicke, Wilkinson, and Peebles were eating lunch in Dicke's office; at the end of the conversation, Dicke famously turned to his colleagues and announced, "Well boys, we've been scooped." A few days later, in Penzias's office, Dicke and Peebles explained to Penzias and Wilson that they had discovered with their Bell Labs antenna what the Princeton team was preparing to search for with their rooftop radio telescope. In May 1965, Penzias and Wilson published a paper in *The Astrophysical Journal* explaining their experimental results, while Dicke and his group published a companion paper interpreting the excess radiation as the signature of the Big Bang. Still, neither group knew about the original predictions of Alpher, Herman, and Gamow regarding the temperature of the CMB.

By 1967, the temperature of the CMB had been measured at nine different wavelengths, from 0.26 centimeters to 49.20 centimeters. All the measurements yielded a consistent answer: the current temperature of the CMB is about 3 K. Over the half century since the original discovery of the CMB, astronomers have measured the CMB signal in nearly every conceivable way, with telescopes hung from balloons, launched into space, and placed nearly everywhere on the surface of the Earth, including Antarctica. The ever-more-refined answer: T = 2.725 K (Chapters Twenty-five and Twenty-six).

The level of accuracy at which the CMB had been measured until about 1990 provided incredibly valuable information about the early history of the universe; however, to obtain an age of the universe from CMB measurements required much more detailed measurements of the CMB, as well as a great deal of additional information about the universe: how much mass it contains, what that mass is made of (normal matter, dark matter, exotic dark matter), how close or far apart galaxies are today, and the past and present rates at which the universe is expanding (dark energy and the accelerating universe). We have all of that information in hand now, and are ready to journey to the edge of modern theoretical and observational cosmology to obtain one more estimate for the age of the universe.

Fickle Fate

The 1978 Nobel Prize in Physics was awarded to Penzias and Wilson for their discovery of the CMB. Dicke was elected to the National Academy of Science in 1973 and would receive numerous, prestigious honors before he retired in 1984. Peebles also has received his share of honors, including the Eddington Medal (1981) from the Royal Astronomical Society and the Henry Norris Russell Prize (1993) from the American Astronomical Society. Gamow, however, died in 1968, his contributions in cosmology all but ignored, though his research in nuclear physics was then and remains now recognized as essential, pioneering work for understanding how stars generate energy. Perhaps if Gamow had lived longer, he might have shared the 1978 Nobel Prize with Penzias and Wilson (though staying alive did not bring a share of the prize to either Alpher or Herman).

The work of Alpher and Herman was all but forgotten until 1977, and even then, despite soon-to-be (1979) Nobel laureate physicist Steven Weinberg describing it in his book *The First Three Minutes* as "the first thoroughly modern analysis of the early history of the universe," few paid them much attention. In 1993, Alpher and Herman did share the Draper Medal, awarded by the National Academy of Sciences "for their insight and skill in developing a physical model of the evolution of the universe and in predicting the existence of a microwave background radiation years before this radiation was serendipitously discovered; through this work they were participants in one of the major intellectual achievements of the twentieth century." Alpher, in an interview in *Discover* magazine in 1999, expressed his frustration at having been so long ignored: "Was I hurt? Yes! How the hell did they think I'd feel? I was miffed at the time that they'd never even invited us down to see the damned radio telescope. It was silly to be annoyed, but I was." It took more than half a century for Alpher's seminal contribution to the field of cosmology to be recognized. He was belatedly awarded the National Medal of Science two weeks before he died in 2007. In writing an obituary for Alpher in the December 2007, issue of *Physics Today* magazine, Martin Harwit, Cornell professor of astronomy emeritus and former director of the National Air and Space Museum in Washington, DC, noted of both Alpher and Herman, that "history will remember their contributions."

CHAPTER 25

Two Kinds of Trouble

Not only was the [COBE] spectrum beautiful to look at, but at one stroke it banished the doubts of almost everyone about the Big Bang theory. For so many decades, the intense combat between the Big Bang and Steady State advocates had continued, and for so many years, a series of small discrepancies between theory and measurement had been explained by ingenious people. Now it was over . . . ."

—John C. Mather, in "From the Big Bang to the Nobel Prize and Beyond," *Nobel Lecture* (2002)

The cosmic microwave background together with the expansion of the universe and the relative amounts of the elements remains one of the fundamental observational pillars of modern Big Bang cosmology. To the limits of experimental accuracy, as measured first by Penzias and Wilson and over the next two decades by all those who worked to improve the accuracy of cosmic microwave background observations, the temperature of the CMB was absolutely identical in every direction (*isotropic*), with a single exception.

This exception, which cosmologists call the *dipole anisotropy*, has nothing to do with the CMB and everything to do with the motion of the solar system at a velocity of 369 kilometers per second relative to the ancient matter that produced the CMB. This motion, which produces a Doppler shift in the observed CMB in the directions toward and away from which the solar system moves, was first detected by Edward Conklin of Stanford University in 1969. Additional measurements over the next decade by several teams nailed down this feature of the observed CMB. Now, when we discuss the isotropy of the CMB,

Map of the CMB temperature
of the entire sky: 2.7 K everywhere

Figure 25.1. When viewed from the Earth, the sky appears as a spherical surface that surrounds us. That surface can be presented in the form of a flat map, just as we might produce a flat map of the surface of the Earth. Using more than two decades of measurements of the temperature of the cosmic microwave background, this map presents the temperature of the CMB sky, as measured prior to 1990. Since these observations showed that the CMB had the same temperature everywhere, this map has no structure.

we are referring to the intrinsic temperature of the CMB after the effect of the motion of the solar system has been taken into account.

The isotropy of the CMB would appear to be consistent with, and even required by, the cosmological principle. That is, if the laws of physics are the same everywhere in the universe, and if the universe looks the same in all directions and is composed of the same constituent elements everywhere, then surely the temperature of the CMB must also be the same in all directions. Yet, seemingly ironically, the initial discovery that the background glow of the universe was apparently perfectly uniform and isotropic, at least to the limiting accuracy of observational measurements, raised enormous problems for cosmologists who could not understand how the CMB could possibly be so uniform and isotropic. We need to look very deeply at this apparent paradox to understand the motivations for the important CMB-related discoveries of the last two decades, which in turn have provided our most accurate estimate for the age of the universe.

Two Problems with the CMB

Two different and completely unrelated problems needed to be solved. The first involved the actual existence of *large-scale structures*—a term used by astronomers to refer to galaxies, clusters of galaxies, superclusters of galaxy clusters, and filaments of superclusters—in the universe:

if the CMB is completely uniform, large-scale structures could never have formed. Galaxies, stars, planets, and people would not exist. We do, however, exist, and so do large-scale structures. It follows that the CMB must have some non-uniform properties (what cosmologists call *anisotropies*), but where in the CMB are they? Cosmologists call this conundrum the large-scale structure problem.

The second problem associated with the CMB is known as the *horizon problem* and concerns the observational evidence that *parts of the universe that appear to be too far apart to have ever been in thermal contact nevertheless have the same temperature.* This property of the CMB could just be a coincidence if two small, distant parts of the universe happen to have the same temperature, but if every part of the universe has the same temperature as every other part of the universe, then we must demand an explanation based on the physics of the universe, rather than on happenstance.

The Large-Scale Structure Problem

If we look out far enough in any direction, we will see galaxies of every type, galaxy clusters and superclusters, filaments, and vast voids between galaxy superclusters and filaments. In this sense and on this spatial scale, the distribution of the large-scale structures in the universe does appear isotropic and homogeneous. But what is quite obvious to any observer who looks up into the sky is that the universe does not look *exactly* the same in every direction and is not composed of *exactly* the same amount of material in every volume of space we might sample.

In one direction we see a star, in another we see the plane of the Milky Way, in another we look off into the dark emptiness of deep space. In one region of space, inside a star, for example, the density of normal matter can be as high as a few hundred million grams per cubic meter (twenty or thirty times denser than iron), while in the nearly perfect vacuum of intergalactic space the density of normal matter can be as low as a single proton per cubic meter. The universe when we observe the components within it that are on the size scale of stars, galaxies, and clusters of galaxies, is not composed of the same material distributed perfectly smoothly throughout space. Instead, the visible mass that is composed of normal matter is clumped. Only when we measure the average amount and kinds of normal matter in volumes as large as hundreds of millions of cubic parsecs does the universe become isotropic and homogeneous. On smaller spatial scales, the universe is anisotropic and heterogeneous.

Since these anisotropies and heterogeneities in the distribution of matter in the universe exist today, they must have existed yesterday and the day before. Conceptually, we can trace them at least as far back in time as the epoch when galaxies and clusters and superclusters of galaxies began to form. But why did these structures form when and where they did? A single word provides the answer to this question: gravity.

Large-scale structures could have formed only if, very early in the history of the universe, certain parts of it contained a little bit more matter and others a little bit less matter than did other parts of the universe. In over-dense locations in the universe, gravity was stronger and as a result the mass already in those locations was able to pull even more matter into its territory, thereby permitting the formation of the precursor objects that have grown into the large-scale structures we observe in the universe today. Without these early density differences, which could have been extremely small, without both clumps and rarefied regions in the primordial distribution of matter, no superclusters of galaxies would exist today, nor would intergalactic voids, clusters of galaxies, galaxies, stars, or planets, and what is worse, there would be no astronomers who could sip cups of tea while discussing and wondering about the existence of these structures.

If these over-dense and under-dense pockets of the universe pre-existed and were the seeds that triggered the formation of large-scale structures in the universe, then some differences in the spatial distribution of matter, which cosmologists call primordial density fluctuations, pre-existed and triggered the formation of these over-dense and under-dense pockets. Almost certainly, these primordial density fluctuations, however small, must have formed from random quantum fluctuations in the first moments after the birth of the universe. The clumps therefore must have existed at the moment the CMB was released and therefore must have left some kind of imprint on the CMB itself. Simply put, the existence of structures in the modern universe places a constraint on the distribution of matter at the earliest epoch in the history of the universe that, in turn, should have caused some parts of the universe to be denser and others more rarefied than average at the moment when the CMB was released. We then must ask, How would these regions of different densities of matter affect the temperature of the CMB? And would these effects be large enough to be measurable?

Solving the Large-Scale Structure Problem: COBE and WMAP

In 1989, NASA launched the satellite COBE, the Cosmic Background Explorer, which began as a suite of mission proposals presented to NASA as far back as 1974. In addition to a telescope, COBE had on board three different instruments designed to measure various aspects of the CMB, one of which was the far infrared absolute spectrophotometer (FIRAS), designed by John Mather. Mather and his team built FIRAS to measure the intensity of the CMB at thirty-four different wavelengths. Within a few weeks after the November launch date, FIRAS had collected a grand total of nine minutes of observations, but this data already provided far more accurate measurements of the CMB than had more than two decades of ground and balloon measurements. At the January 1990 meeting of the American Astronomical Society, John Mather presented the fruit of these first nine minutes to a crowd of more than 1,000 hushed and eager-to-know astronomers. Mather announced that COBE had pinned down the average temperature of the CMB to 2.735 K, with an error of only 0.06 K. It was an incredible accomplishment, and earned the COBE team a spontaneous and unprecedented standing ovation. The final FIRAS results, reported in 2002 by Mather, refined the measurements of the average temperature of the CMB to 2.725 ± 0.001 K.

The goal for the COBE science team, though, was to measure both the average temperature and any temperature differences in the CMB,

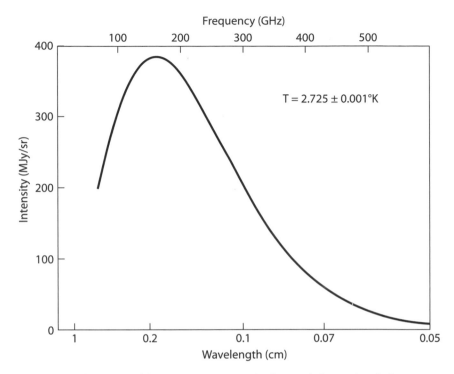

Figure 25.2. Spectrum of the cosmic microwave background obtained with the instrument FIRAS on the COBE satellite. The errors on the thirty-four different measured data points are so small that they cannot be seen under the line drawn for a blackbody with a temperature of 2.725 K. Image courtesy of NASA.

which it was hoped would be more readily detectable from COBE than from the balloon-borne detectors of the 1960s through the 1980s with their very limited spatial resolution and limited sensitivity to small temperature differences. Imagine a 1960s weather satellite looking down at North America. The satellite measures an average temperature for the whole continent; yet, when viewed with more spatial detail, with a better, future-generation satellite, we find that Texas is hot, Nebraska is less hot, North Dakota is cool, and Manitoba is very cold. No single location is "average." Our weather satellite observations will reveal the average unless we have a satellite capable of seeing sufficiently fine *spatial* detail. In addition, even if our satellite is able to separately measure the temperatures in Texas and Manitoba, if the

Figure 25.3. World map, showing how the three-dimensional surface of a sphere is depicted as a flat map. The large circle over North America illustrates how a single measurement of the temperature of the Earth, obtained with low spatial resolution, would measure the average temperature over a wide range of latitudes. Such a measurement would not be able to indicate the wide range of actual temperatures. Observations made with higher spatial resolution (small circles) would be able to distinguish the high temperatures in Texas from the much lower temperatures in Canada.

temperature in Texas is 300 K while the temperature in Manitoba is only 299.9 K and if our measuring device is only accurate to the nearest degree, we would be unable to detect the *temperature difference* between these two locations on the continent. Similarly, in order for astronomers to detect any temperature differences in a microwave map of the sky, they needed a telescope that could be pointed at small enough regions of the sky and that could detect very small temperature differences.

The differential microwave radiometer (DMR) on COBE, designed by George Smoot, was built to measure differences in the temperature of the CMB in any two directions separated by an angle of 60 degrees—if any were there. It did its job. The DMR team detected fluctuations at the level of 36 millionths of one degree (barely one thousandth of one percent of the CMB temperature). After a quarter century of effort, astronomers finally knew the absolute temperature of the CMB: T_{CMB} = 2.725 K. And after an equally long pursuit, they knew that the

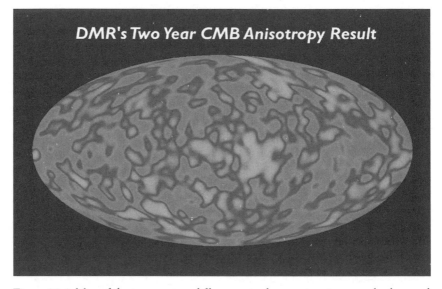

Figure 25.4. Map of the temperature differences in the cosmic microwave background measurements made by COBE. Regions inside black contour lines are as much as 0.0001 K cooler, while gray regions without black contour lines are as much as 0.0001 K warmer than 2.725 K. Image courtesy of NASA.

absolute temperature of the CMB was not identical in all directions. The differences in temperature (ΔT) from one part of the sky to another were very tiny, but thanks to the DMR instrument and to the analysis of its data by George Smoot's team, they were now a known quantity: $\Delta T = \pm 0.000036$ K. For their accomplishments in measuring so accurately the average temperature of the CMB and mapping the tiny temperature differences in the CMB across the sky, Mather and Smoot shared the Nobel Prize in Physics for 2006.

COBE, however, did not have the spatial sensitivity to measure cosmic detail on scales as small as clusters of galaxies. The smallest detail COBE could see was about seven degrees across (equivalent to a circle on the sky with a diameter about the size of a fist held at arm's length). As a result, any regions of space that were hotter than about 2.7251 K or colder than about 2.7249 K but smaller in their angular extent than 7 degrees across could not be detected by COBE; any such higher and

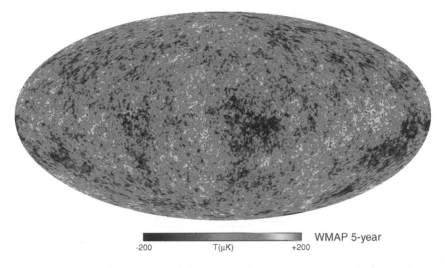

WMAP 5-year

-200   T(μK)   +200

Figure 25.5. Map of temperature differences in the cosmic microwave background measurements made by WMAP. Dark regions are as much as 0.0002 K cooler, while whitest regions are as much as 0.0002 K warmer than 2.725 K. Image courtesy of NASA and the WMAP science team.

lower temperature fluctuations in the CMB from small angular areas were beyond the limited spatial resolution capabilities of COBE.

The follow-up mission to COBE, directed by Charles Bennett, was a satellite known as the Wilkinson Microwave Anisotropy Probe (WMAP). It was launched in 2001. WMAP has a spatial resolution of less than 0.3 degrees across, thirty-three times better than COBE, and so is able to see much finer details in the spatial distribution of the CMB. According to the WMAP results, the temperature fluctuations in the CMB on a 0.3 degree spatial scale are two ten-thousandths of a degree (0.0002 K), six times larger than those seen by COBE: T_{CMB} = 2.725 and ΔT = ± 0.0002 K. Thus, COBE and WMAP appear to have answered the first question raised by the apparent smoothness of the CMB: How could it be the same in all directions despite the presence of large-scale structures in the universe? The answer is simply that it is not. Temperature differences as large as one thousandth of a degree exist between the hottest and coldest spots.

The Horizon Problem

The second problem raised by the spatial uniformity of the CMB, the horizon problem, was not solved by the presence of temperature fluctuations of one or two ten-thousandths of a degree. Instead, the solution came instead through a radical idea called *inflation*. But before we solve this problem, let's restate it.

We know that the ratio of the size of the universe today to the size of the universe when the CMB was released is the same as the ratio of the temperature of the universe then (3,000 K) to the temperature of the universe now (2.725 K). That ratio is almost exactly 1,100. We therefore know that the universe today is larger by a factor of 1,100 than it was at the moment the CMB was released.

Now think about what astronomers do when they set out to observe the CMB. They build or select for use a telescope, point that telescope in a selected direction in the sky, and measure the temperature of the CMB. The result: T = 2.725 K. The CMB photons that arrived at this telescope from the selected direction are, to restate the obvious, photons. Photons are particles of light that travel at the speed of light. They have been traveling through the universe since the CMB was released, 380,000 years after the first moment of the Big Bang. These photons have arrived at the Earth having traveled for almost 14 billion years across a distance through space of almost 14 billion light-years.

Twelve hours later, after the Earth has rotated 180 degrees, our intrepid astronomer points her telescope in the exact opposite direction in the sky (CMB observations can be made during the nighttime or daytime) and measures the temperature of the CMB. The result is unchanged: T = 2.725 K. But the CMB photons measured in the second set of observations also have traveled at the speed of light for almost 14 billion years and traversed nearly 14 billion light-years of space, but from a different direction. The two sets of photons have come from parts of the universe that are now at least 28 billion light-years apart, yet they have the same temperature, as would any photons from any two directions in which we might look, to an accuracy of better than one part in one hundred thousand.

Keep in mind that the universe has expanded by a factor of 1,100 in

14 billion years. Those two locations that are today separated by at least 28 billion light-years were separated by only about 25 million light-years when the CMB photons were released. Yet the universe was only 380,000 years old at that moment in time. How can two regions of space separated by 25 million light-years have had the same temperature when the universe was only 380,000 years old? To paraphrase astronauts John Swigert, Jr., and James Lovell, who in 1970 were reporting a major malfunction on the Apollo 13 spacecraft to mission control, Houston, we have a problem. And the problem involves the limited speed at which information (including about temperature) can be communicated across the universe.

We need to think about temperatures and how objects come to have identical temperatures. Consider a baked potato, fresh out of the oven, which has been set on a plate that is placed on a table. The potato, once out of the oven, is a hot object placed in a much cooler environment. What happens of course is that over some short period of time, the potato cools. Where does the heat from the potato go? Into the surrounding air. The air heats up while the potato cools off. And what happens to an ice cube dropped into a cup of hot tea? Given enough time, the ice cube warm ups and melts while the tea in the cup cools. Any cold object placed in a hot environment will warm up while the environment simultaneously cools off.

All objects placed *in thermal contact* with other objects *for a sufficient amount of time* will thermally equilibrate; that is, they will exchange heat until they have the same temperature. To do this, the hot and cold objects communicate with each other. They communicate through collisions of particles and through those collisions, low-energy (lower temperature) particles gain energy at the expense of high-energy (higher temperature) particles. This is why, on a cold winter's day, when a door to the outside is opened and then quickly shut again, the cold gust of air that sweeps into the room is felt only briefly. Very quickly, collisions of colder air molecules from outside with warmer air molecules from within the room blur the differences in temperature, bringing all the air molecules into thermal equilibrium.

If the colliding particles are air or water molecules, heat is a measure of their speeds. Faster-moving molecules slow down after collisions with slower-moving particles while those slower moving particles speed up. If, however, the colliding particles are photons, which all move at the same speed, the scattering of photons off electrons cannot change the speeds of the photons, but the collisions do transfer energy to the electrons. In turn, when a photon transfers energy to an electron, the photon is either redshifted (if it loses energy) or blueshifted (if it gains energy), and through these collisions, as well as through electron-electron collisions, both the photons and the free electrons come into thermal equilibrium.

In order for the CMB to have the same temperature in all directions as measured by earthbound observers today, 14 billion years ago every location in the universe must have been in thermal contact with every other location in the universe for long enough that any pre-existing temperature differences would have been smoothed out though collisions of particles and photons. Yet, we were able to calculate that two regions of space that had exactly the same temperature when the universe was only 380,000 years old were separated at that time by 25 million light-years. Since no information can travel *through space* faster than the speed of light, these two regions of the universe appear to have been *causally disconnected*. The particles and photons in these two regions of the universe were unable to communicate with each other; no collisions could occur between particles or photons in one region and particles or photons from the other region. Neither could say to the other, "I'm hot, you're cold, let's equilibrate."

The solution to this apparent paradox lies in special relativity. Specifically, in the fact that special relativity precludes the movement of particles, including photons, *through space* at speeds greater than the speed of light but does not prevent space itself from expanding at speeds greater than the speed of light (superluminal expansion). Imagine that shortly after the universe began, when it was so small that all parts of it could exchange information through collisions of particles, this entire (albeit very small) universe had reached thermal equilibrium. It then inflated at hyperspeed—much faster than the speed of light—for an extraordinarily brief period of time. After that nearly infinitesimally short inflationary epoch, most parts of the universe found themselves too far apart to communicate with most other parts of the universe, too far apart to say, "I'm hot, you're cold, let's exchange energy." If one part of the universe were colder than another, it would have to stay that way. But if all parts of the universe had exactly the same temperature before the inflationary epoch, they would necessarily still have identical temperatures when inflation ended. After the epoch of superluminal expansion ended, the almost exactly isothermal universe would cool as it continued to expand, and since the universe must expand uniformly so that the cosmological principle is not violated, all parts of the universe would necessarily cool at the same rate. Therefore, every part of the universe must have released CMB photons at the same moment and with the same characteristic temperature, despite being separated at the moment of release by distances vastly greater than the distances at which they could be in thermal contact. This idea, proposed and dubbed *inflation* by Alan Guth in 1981, describes an epoch in Big Bang history when the universe inflated at a speed faster than the speed of light.

The inflation theory allows us to think of the universe as having three distinct historical phases, which we will refer to as before inflation, during inflation, and after inflation. Before inflation, all regions of the universe were in causal contact and, as a result, both the density and temperature of the universe was almost uniform throughout; the nearly constant density of the universe before inflation preserved thermal equilibrium throughout the young universe and led, 380,000 years later, to the emission of light from regions throughout the entire universe that was of uniform temperature. During the inflationary epoch, the diameter of the universe expanded from a size roughly a billion times smaller than the diameter of a proton to about the size of a softball. This increase in volume by a factor of about ten to the fiftieth power (10^{50}) occurred when the universe was only ten to the minus thirty-five seconds old (ten billionths of one billionth of one billionth of one billionth of a second) and lasted until the universe was about ten to the minus thirty-four seconds old (one hundred billionth of one billionth of one billionth of one billionth of a second). Because the material in the universe was distributed so uniformly at the onset of inflation, and because all parts of the universe inflated uniformly, after inflation the material in the universe remained as uniformly distributed as it had been before, despite the fact that most parts of it were now beyond causal contact with most other parts (that is, the distance between them was greater than the speed of light multiplied by the age of the universe).

The small fluctuations in the temperature of the CMB detected by COBE and WMAP solve the large-scale structure problem. Inflation solves the horizon problem by permitting the entire universe to have been causally connected before the inflationary epoch; density differences in the pre-inflationary universe that were extremely small compared to the average density were preserved by inflation and those incredibly small fluctuations generated the minute temperature differences registered by COBE and WMAP. With these two problems solved, we can return to our original goal: we will use the map of the CMB produced by the WMAP team to estimate the age of the universe.

···CHAPTER 26···

The WMAP Map of the CMB
and the Age of the Universe

The age of the best-fit universe is t_0 = 13.7 ± 0.2 Gyr. Decoupling was $t_{dec} = 379^{+8}_{-7}$ kyr after the big bang at a redshift of z_{dec} = 1089 ±1 . . . This flat universe model is composed of 4.4% baryons, 22% dark matter, and 73% dark energy.

—Charles L. Bennett, in "First-Year Wilkinson Microwave Anisotropy Probe (WMAP) Observations: Preliminary Maps and Basic Results," *The Astrophysical Journal Supplement Series* (2003)

We're zeroing in on our goal now. In fact, with the WMAP map in our toolkit, we have just about everything we need to calculate an age for the universe. But first we must learn how to read the map. We need also to keep in mind that any age estimate based on the WMAP map will be model-dependent; that is, it will only be as reliable as cosmologists' understanding of dark energy and their calculations of the amounts of dark and normal matter that exist in the universe. What we want to find out is whether the age arrived at using data from WMAP will be consistent with and supportive of the ages we have derived from the three independent methods described in earlier chapters. Recall that these methods were based on white-dwarf cooling times, the ages of star clusters, and the physics of the expanding universe.

At the most fundamental level, the spot pattern in the WMAP map is the signature of temperature differences that existed in the gas that filled the universe when it was 380,000 years old. While any single spot represents a region with a temperature different from the temperatures of the spots that surround it, some spots are small and others are large and the map shows more small spots than large spots. The temperature differences were the result of pressure differences in the gas. Those pressure differences were caused by sound waves that were traveling through the gas. The relative amounts of exotic dark matter and normal matter and the laws of physics were the primary controlling factors in shaping and driving these sounds waves.

In "reading" the map, cosmologists reverse engineer it. That is, they use the characteristics of the map to determine the relative and absolute amounts of exotic dark matter, normal matter, and dark energy in the universe. Since both kinds of matter act to slow down the expansion of the universe, while dark energy has the opposite effect, our knowledge of the relative and absolute amounts of these three components of the universe allow cosmologists to calculate the length of time needed for the universe to expand to its current size.

The actual nuts and bolts of the map-reading process involve advanced mathematical techniques and produce a plot called a *power spectrum*. The profile of the power spectrum reveals the secrets encoded in the map. Although we will not dig into the mathematics that the WMAP team used to analyze the map of the cosmic microwave background, we can nonetheless understand the basic concepts involved.

Color Blindness Charts and the WMAP Map

The first thing we need to do is understand how we can pull information about the physics of the early universe from a map consisting of spots of different sizes, each of which represents a volume of space in the early universe with a temperature just a bit different from the temperatures of the volumes of space (other spots in our map) that surround it. We'll start with colorblindness charts, which are close visual analogs to the WMAP map. If we can understand how information about colorblindness is extracted from colorblindness charts, we can understand, in broad outlines, how cosmologists read maps of the cosmic microwave background. It's a short cut, but it will save us having to spend several years in gradual-level mathematics classes.

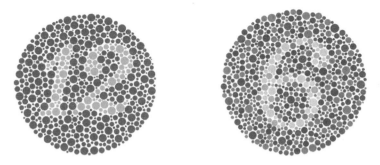

Figure 26.1. Two grayscale versions of Ishihara colorblindness charts. Left, the number "12." Right, the number "6." In full-color versions of these charts, colorblind readers would see a random assortment of dots of different sizes and colors in the panel on the right.

In 1918, Japanese ophthalmologist Shinobu Ishihara created visual charts for testing colorblindness. Each of these charts is an almost circular shape, with each circle made from a large number of small, colored spots. Some of the spots are small, some are large; and they come in a variety of colors. Embedded in the otherwise apparently random distribution of colored spots is a pattern, typically a number or shape. When a person who is colorblind looks at a typical Ishihara chart, he (7 percent of men but only 0.4 percent of women are red-green colorblind) sees no pattern at all, just a random splattering of spots of different sizes and colors; however, a person with a strong sense of colors and shades will see the embedded patterns in the distributions of dots on the charts.

The trick to seeing the patterns is having the right tools. We detect color using pigments in the cones in our retinas, so in order to see colors of light we need a set of genes that properly codes for the production of these pigments. If our genes code for the wrong pigments, the cones will lack sensitivity to the full range of colors and we will be blind to some or all of the information embedded in the Ishihara charts.

Like a circular Ishihara colorblindness chart, the oval COBE map of the cosmic microwave background shows a distribution of spots with a range of sizes; although the spots are color-coded according to how different their temperatures are from the average CMB temperature of

2.725 K, the information cosmologists will pull out of the COBE map (or the more sensitive WMAP map) will be based exclusively on the pattern of shapes, not on colors.

The fundamental laws of physics enable cosmologists to predict the range of possible patterns—not all patterns are possible—that could be in a map of the CMB. To better understand this restriction, imagine you were asked to draw a map of the surface of the Earth. You are told that the following "law" applies to your map: the surface includes both continents and oceans. Given this law, many variations are possible in the relative proportions of land area to ocean area and in the shapes of the continents and oceans. Two maps, however, can be excluded because they violate the law: 1) a map with only oceans and no continents; and 2) a map with only continents and no oceans. You then discover a second law that further limits the design of your map: the ratio of ocean area to continental area is three to one, and the smallest land mass covers 900 square kilometers. Although many surface maps remain possible, this second law eliminates an enormous number of possible variations that would violate its strictures: any map with too much or too little continental area or with small islands is ruled out.

Using the mathematical tools of their trade, cosmologists look for patterns in the spatial and size distributions of shapes on the WMAP map. They then compare the observed patterns with patterns that can be predicted from the laws of physics or from other known constraints about the universe (e.g., the amount of mass, the amount of dark matter, the amount of dark energy, the relative amounts of hydrogen and helium). By identifying the best match between the observed pattern and physically plausible patterns, cosmologists have been able to identify the physical processes that in the early universe generated the observed pattern of the CMB. Ultimately this modeling process provides an estimate of the age of the universe as well.

Pattern Recognition

We now understand how cosmologists look for patterns in a map of the CMB. We also understand that, just as Dr. Ishihara used his own rules for colors and numbers to create and impose structure on his color-blindness charts, the physics of the early universe imposed structure (equivalent to the pattern and distribution of spot sizes) in the form of density variations among the particles and photons in the early universe. Because these density variations existed when the CMB was released, they in turn imposed temperature variations on the CMB. We see these temperature variations in the pattern of spot sizes when we make maps of the temperature of the CMB.

Now let's think through this logical chain of ideas in reverse. We measure the temperature variations in the map of the CMB in the form of spots of different sizes (each spot is characterized by a temperature that is different from the temperatures of its neighboring spots). These temperature variations exist in the CMB today because of density differences in the distribution of the matter that released the CMB photons. Since the CMB photons were released 380,000 years after the beginning of the universe, those density differences existed at the moment when all the free electrons combined for good with all the free protons. Those density differences existed at that moment in time because of the behavior of particles and photons in the epoch that immediately preceded the release of the CMB photons. The behaviors of those particles during all earlier time periods were governed by the laws of physics for charged particles and photons in the dense, hot environment that was the newborn universe. Therefore, some process, acting through the laws of physics at the earliest moments in the history of the universe, forced the particles and photons to move in a particular way, and that process imposed structure in the form of a pattern of density differences onto the universe.

The Power Spectrum for a Kitchen Tile Pattern

The structure imposed on the physical universe during the earliest years of universal history is preserved in the pattern of the CMB. When cosmologists use the tools of their trade to "read" and decode the pattern that is encoded in the map of the CMB, their preferred way to present their results is through a plot they call the "power spectrum of the CMB." Our goal therefore is to examine and understand that plot; but in order to reach that goal we first must understand what a *power spectrum* is.

The most basic way mathematicians identify spatial patterns is to look for phenomena that repeat across a distance or throughout an area. Here are two examples:

- railroad ties are spaced about 21 inches apart. That spacing can be explained as a spatial frequency, which describes how often a phenomenon occurs across a certain distance. In this case, the railroad ties have a spatial frequency of about 3,000 railroad ties per one mile length of track.
- bricks for building houses and walls have a standard height of 65 millimeters with a 10-millimeter layer of mortar between bricks. This produces a spacing of 75 millimeters between the centers of the bricks, from the bottom to the top of a brick wall, or a spatial frequency of 13.33 bricks per meter.

Now let's look at a more complicated, closer-to-home example: the interior design plans for my new kitchen. In my kitchen, I want to use one-foot (twelve-inch) tiles on the floor, one-third-foot (four-inch) tiles for the counters, and one-twelfth-foot (one-inch) tiles on the walls above the counters. With these tile sizes, the room will have three different spatial frequency patterns: one tile per square foot on the floor, nine tiles per square foot on the counters, and 144 tiles per square foot on the backsplash. If the floor covers 300 square feet, the counters 50 square feet, and the backsplash 20 square feet, I will need 300 twelve-inch tiles, 450 four-inch tiles, and 2,880 one-inch tiles to complete the project.

Because the 12-inch tiles cover 300 square feet while the 4-inch tiles cover only 50 square feet and the 1-inch tiles cover only 20 square feet, the number of tiles of each size do not measure the visual impact of each tile size in my kitchen. Instead, a better measure of the visual impact of the tiles would be the square footage of the kitchen covered by each tile size. We will call the square footage the *power*, and if we make a plot of the power at each spatial frequency, we will have a plot

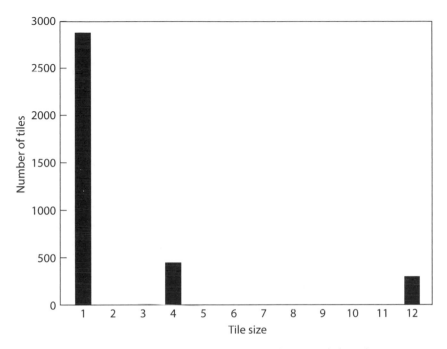

Figure 26.2. Graph showing the number of tiles of each size needed to tile my kitchen: 2,880 1-inch tiles, 450 4-inch tiles, and 300 12-inch tiles.

called a *power spectrum*. In this particular kitchen, the power spectrum shows that most of the power was expended at the smallest spatial frequencies (300 units of power is seen for one-tile-per-square-foot 12-inch tiles), a moderate amount of power is evident at the intermediate spatial frequencies (50 units of power for the nine-tiles-per-square-foot 4-inch tiles), and very little power shows up at the highest spatial frequencies (20 units of power for the 144-tiles-per-square-foot 1-inch tiles).

Different kitchen designs, with different sizes of floors, counters, and backsplashes, and perhaps different choices for tile sizes, would naturally have unique patterns of spatial frequency versus power. We might even imagine what a power spectrum would look like for the tile design for an entire house, for a medieval church, or for a modern city. Each would show a suite of tiles-per-square-foot needs, from sidewalks (with 6-foot-by-4-foot segments, the spatial frequency for sidewalk tiles would

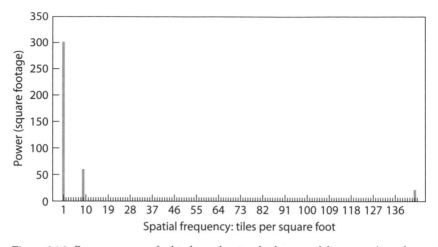

Figure 26.3. Power spectrum for kitchen tiles, in which most of the power (i.e., the square footage covered by tiles of each size) is at low spatial frequencies (big tiles): 300 units of power for the one-tile-per-square-foot tiles, 50 units of power for the nine-tiles-per-square-foot tiles, and 20 units of power for the 144-tiles-per-square-foot tiles.

be very low, only one-twenty-fourth of a tile per square foot) to intricate mosaic fountains (with thousands of tiny tiles per square foot, the spatial frequency for the mosaics would be enormous). In turn, the power per spatial frequency would depend on the total areal coverage of the sidewalks in the city and the surface area for the fountains in the parks.

Now, instead of planning to build this kitchen, for which we *a priori* know how much power will be needed at each planned spatial frequency for our tiles, let's imagine that we are visiting a new home that has already been finished. We fall in love with the design and decide we want to build another kitchen with the identical patterns. Our job is to make a map of the kitchen, determine the various spatial frequencies of the tiles, and determine the spatial frequency versus power pattern for all possible spatial frequencies that were part of the design plan of tiles for this particular kitchen.

When we have completed our job, we will have uncovered some of the fundamental architectural principles that underlie the blueprints for this house, and (if we wish) we will be able to purchase the neces-

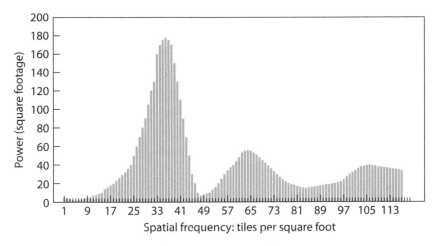

Figure 26.4. Power spectrum for an entire house. In this house, tiles of every size, from one tile per square foot down to more than 100 tiles per square foot, have been used, but larger areas (i.e., power) of the house have been covered in tiles of certain sizes than other sizes. For example, very little power is found in the very largest tiles (one tile per square foot), an increased amount of power is found in intermediate-sized tiles, while the greatest amount of power seen in two-inch-by-two-inch tiles (thirty-six tiles per square foot). This house also has a large amount of power (a secondary peak) in 1.5-by- 1.5 inch tiles (sixty-four tiles per square foot).

sary supplies for building another kitchen exactly like the one we covet. What if we measured and computed the power spectrum and found lots of power at spatial frequencies of 144 per square foot (one-inch tiles), 36 tiles per square foot (two-inch tiles), 16 tiles per square foot (3-inch tiles), 5.76 tiles per square foot (5-inch tiles), 2.94 tiles per square foot (7-inch tiles), and 1.19 tiles per square foot (11-inch tiles), but no power at any other spatial frequencies. We could reasonably conclude that the architect had imposed an idiosyncratic rule requiring that all tiles in this house must have lengths, measured in inches, that are prime numbers smaller than 12. Just as the power spectrum for this kitchen can reveal the architectural principles that guided the construction of this house, our analysis of the power spectrum of the CMB has the potential to reveal information about the underlying physical principles (e.g., the amount of dark matter, the importance of dark energy) that guide the evolution of the universe.

One way, a very inefficient and time-consuming way, to obtain the power spectrum of this new house would be to painstakingly chart and count, one by one, every tile in the house. A physicist would use a more efficient but equally precise approach; she would take pictures of the kitchen and apply pattern recognition techniques to find the many tile sizes and determine how the power is distributed among the many patterns. This is exactly what cosmologists have done with maps of the CMB. They search for patterns that might not be obvious to our eyes, not because we are colorblind (and remember, the patterns cosmologists are looking for have to do with the distribution of sizes, not colors) but because many patterns are so subtle that our eyes do not easily recognize them until after they are pointed out to us.

The Power Spectrum of the CMB

Cosmologists have obtained the power spectrum of the CMB from the WMAP map. That power spectrum is presented in the form of power as a function of a parameter cosmologists call the *multipole moment* (symbolized by "*l*", which is plotted along the x-axis in Figure 26.5). The multipole moment parameter bears some similarities to but is also a bit different from the tile-size parameter we used in our power spectra of kitchens and houses because the multipole moment represents both the size of the temperature spots on the sky and how often those spots occur around the circumference of a circle. Imagine that you manage a restaurant with only circular tables in the dining area. If you wish to seat four people at one of these tables, each chair can be wide, but if you wish to seat a party of twenty at the table, the chairs will necessarily be quite narrow. Eventually, to help your employees know how to set up the dining room each evening based on the known reservation list, you come up with a parameter we'll call the "dining-room-table-and-chair moment" (*d*) that encodes information about both the chair sizes and the numbers of chairs that will encircle each table. You might have ten parties of six that evening (lots of power in the dining-room-table-and-chair moment for $d = 6$) but only one party of twenty that evening (very little power for $d = 20$). The combination of all the reservations would yield your power spectrum for your restaurant. With

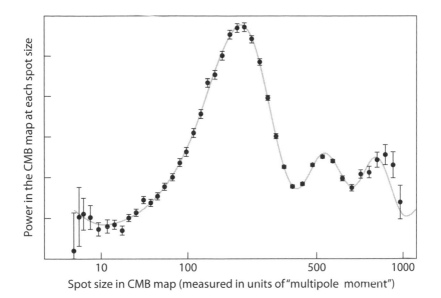

Figure 26.5. The power spectrum of the cosmic background radiation obtained by the WMAP team from five years of data. The values plotted along the horizontal axis are equivalent to spot size in the CMB map, with small multipole moment values corresponding to large (many degrees across) spots and large multipole moment values corresponding to small (fractions of a degree) spots. We can think about the multipole moment as being proportional to the number of spots that could be fitted around the equator of a sphere. The the total amount of power in the map at each spot size is plotted along the vertical axis. The line through the data points is the WMAP team's model fit for a universe with an age of 13.69 billion years. Image courtesy of the WMAP science team.

the multipole moment parameter, we consider how to fit a set of spots around a circle at the equator of a sphere. If we fit only a few spots around that circumference, those spots could be large; if we wanted to fit more spots around that circumference, the spots would have to be smaller; hence the multipole moment is sensitive to both spot size and how often they appear (if l = 2, the spot pattern repeats about every 90 degrees, if l = 3, the spot pattern repeats approximately every 60 degrees, and if l = 4, the spot pattern repeats nearly every 45 degrees). The CMB power spectrum looks very much like the power spectrum for our entire house. Very little power exists in the CMB in the largest spot sizes (the smallest values of the multipole moment); instead, most of the power is in spot sizes ranging from about 2 degrees (l of about 60) to about 0.5 degree (l of about 400), with the dominant spatial frequency corresponding to an angular size of about 0.60 degrees (l of about 220). The CMB power spectrum also shows two additional power peaks at about multipole moments 540 and 800 (angular sizes of about one-third and one-fifth of a degree, respectively).

Just as the power spectrum for our house contains a great deal of information about the structure of our house, the considerably more complicated power spectrum of the CMB hides a great deal of information about the physics of the universe. All we have to do is learn how to decode it.

Sound Waves Create the CMB Pattern

Let's summarize what we know so far. We know that a pattern exists in the CMB map. We now have the power spectrum for that pattern. We also know that the pattern in the CMB map represents temperature differences in the heat emitted by different regions of the universe at the moment in time when the CMB was emitted. These temperature differences were generated because the very young universe had some regions that were dense and others that were more rarefied. The existence of both dense and rarefied regions is not a surprise, since without these density fluctuations gravity would have been unable to trigger the formation of the large-scale structures (galaxies and clusters of galaxies) that exist in the modern universe. Now, in order to obtain an estimate for the age of the universe from this power spectrum, we need to dig deeper into the physics that generated these density fluctuations.

The physical behavior of the early universe is a product of the interplay of a number of factors: the total amounts of normal matter and exotic dark matter within the early universe, the contribution of dark energy to its total energy content, and the total numbers of photons and neutrinos within it. All of these aspects of the early universe contribute to making the power spectrum of the CMB look the way it does. Our next step, therefore, will be to try to understand the broad or narrow range of plausible conditions that could produce this power spectrum.

Before the CMB photons were released, the young universe was filled with ionized particles of normal matter (almost entirely hydrogen and helium nuclei and electrons), in addition to photons, neutrinos, and exotic dark matter particles. By sheer number, most of the particles were photons. While the weakly interacting neutrinos and exotic dark matter particles almost never collide with other particles, the photons and normal matter particles could travel only short distances before ricocheting off of other similar particles. When particles collide and bounce off of one another, they are acting as a gas and they generate pressure. If the number of collisions among these particles is large, the gas pressure is high. The physical laws that govern the behavior of gases and those laws that describe the interactions between ionized particles and photons are very well known, so cosmologists understand reasonably well the properties of this gas at all times before the CMB photons were released.

One of the properties of gases is that a disturbance in one region of space, whatever the cause, can increase the pressure in that region, making the pressure in the disturbed volume of space higher than the pressure in the space surrounding it. A high-pressure region surrounded by lower-pressure regions is unstable: the high-pressure region must expand. When a high-pressure region expands it decompresses; it also does work on the surrounding low-pressure regions, compressing them and thereby driving up their pressure. When the pressure in the expanding volume drops to a pressure that matches the rising pressure of any and all surrounding volumes, it should stop expanding since the pressures are now in balance; however, while the pressures are in balance, the gas particles are moving and cannot stop instantaneously. As a result, the decompressing gas volume overshoots the moment at which the pressures are equal, expands too much, and becomes more rarefied than the surrounding gas, which itself is now overcompressed. The newly overcompressed regions, now being at high pressure, are able to push back and stop the expansion. Now the situation is reversed but again we have a high-pressure volume surrounded by low-pressure volumes. Again, the high-pressure region must decompress by

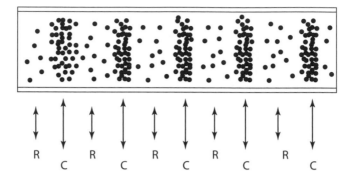

Figure 26.6. A pressure wave in a tube. High-pressure regions (C = compressed) alternate with low-pressure regions (R =rarefied). When the compressed regions expand, they will compress the rarefied gas regions. An initial disturbance at one end of the tube will compress the gas at that end. This disturbance will progress down the tube as a pressure (sound) wave.

compressing the lower-pressure regions next to it. Thus, any small disturbance will generate a pressure wave—an endless cycle of compression followed by decompression followed by compression, and so on and on—that will propagate through the gas. Since a compressed volume of gas heats up and a rarefied volume of gas cools off, pressure waves necessarily also create regions of hotter (higher pressure, higher density) and cooler (lower pressure, lower density) gas.

Pressure waves traveling through any kind of matter capable of compression and rarefaction are known as sound waves because our ears can detect their presence in the air. But sound waves, of course, can exist even if no one with ears and eardrums is around to hear them.

Gravity also plays a subtle but important role in the propagation of sound waves in the early universe. A high-pressure region is denser and therefore contains more mass than a low-pressure region of the same size. The force of gravity, which is generated by both normal and exotic dark matter, is therefore stronger in high-pressure regions than in low-pressure regions. The greater gravitational force acts to squeeze the high-pressure volume even harder than the pressure wave would squeeze it by itself, thereby enhancing its compression, making the compression phase of the cycle stronger and deeper than it would be without the influence of gravity. The depths of the compressions therefore encode information about the total amount of mass (normal plus exotic dark matter) in the universe.

Radiation pressure (generated by collisions between photons and other particles) also increases with increasing density; however, radiation pressure acts in the opposite way to mass: it pushes back against gravity, slows down the compressions, and even pushes back to generate expansion. Since more mass means higher density, which generates greater compression and thus greater pressure and stronger push-back against gravity from radiation pressure, the importance of and push-back from radiation pressure plays a greater role in the physics of the early universe if lots of mass existed than if little mass existed and the role of gravity was minimal. It follows that when pressure finally stops the compression and forces the compressed volume to begin to expand outward, the rate of outward expansion is more forceful in a universe with lots of mass than it would be in a universe with little or no mass, and so the overshoot of this expanding volume is greater than it would be in a universe with little or no mass. Yet, the dark matter, whose gravity contributed to increasing the strength of the compressions, does not help generate additional outward pressure since it is weakly interacting (i.e., collisionless), and pressure depends on collisions. As a result, the dark matter enhances the compression phase of the cycle but hinders the expansion phase.

The theory of how pressure or sound waves behave (which is known as the *theory of acoustic oscillations*) provides cosmologists with the tools for interpreting the CMB measurements. To understand how the theory of acoustic oscillations applies to the universe as a whole, let's think of the universe as a flute. A flute is a container, a volume of space, filled with gas. Any initial disturbance, like that caused by a flautist sending a puff of compressed air across the opening of her flute, generates a sound wave that oscillates in the cavity of the flute. Similarly, any disturbance anywhere in the young universe would send sound waves outward through the universe, emanating from the location of the disturbance. The size of the musical instrument plays a large role in determining the range of sounds that can be produced with that instrument;

because a tuba is made from a longer pipe than a trumpet, the tuba can support longer wavelength (lower sound) waves than can the trumpet while a baritone saxophone produces lower sounds than does an alto saxophone.

The flute, like any musical instrument, can produce a range of sounds, but each note is not a single sound wave that escapes the flute and arrives at our ears. Instead, each note is composed of many sets of vibrations, known as the fundamental tone and overtones. The fundamental tone and overtones emerge from different wavelength waves generated inside the musical instrument, with the overtones having wavelengths of one-half, one-third, one-fourth, and one-fifth (and beyond) of the fundamental. These many wavelengths are determined by the size and quality of the instrument. The difference between a cheap violin and a Stradivarius, or a dime store guitar and a Gibson, are the wealth of overtones generated and sustained in the better-made instrument.

Since wavelengths can be restated in terms of the frequencies for these sound waves, we can think of our flute as producing sounds that have a range of frequencies. A skilled sound technician can use the tools of her trade (i.e., a microphone attached to a sound recorder linked to a computer with software designed to analyze the spectrum of sounds) to make measurements of the frequencies of sound emerging from an orchestra and the amount of power generated by the musicians and their instruments at each frequency. Such measurements are no different from the way an interior designer (trained as a physicist) would measure the spatial frequencies of the ceramic tiles and the power spectrum that describes their use in my kitchen.

Sound Waves in the Early Universe

The CMB map shows spots of different sizes and patterns of different length scales. These spots and patterns have their origin in the sound waves traveling through the early universe; those sounds waves were triggered by disturbances in the very early universe; and the fundamental tone and overtones of those sound waves were determined by the

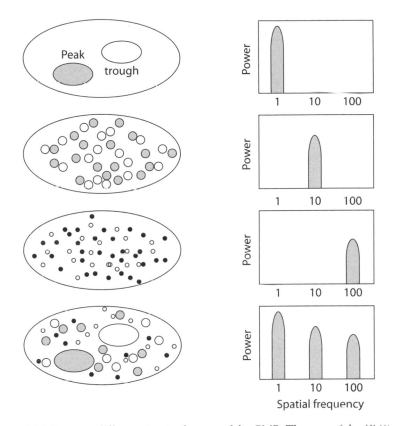

Figure 26.7. Power at different sizes in the map of the CMB. The map of the CMB shows big sound waves (top, left) that appear as peaks (high-temperature regions) and troughs (low-temperature regions) on the CMB map and produce power at a low spatial frequency (top, right). The CMB map also shows intermediate-sized (top-middle, left) and very small-sized (bottom-middle, left) sound waves, each producing power at a unique spatial frequency. The actual map of the CMB is a composite of waves of many different sizes (bottom, left), with a different amount of power at several spatial frequencies, as seen in the power spectrum.

physics of the newborn universe. The power spectrum generated from analyzing the CMB map indicates that the entire universe was awash with pressure waves of different wavelengths (the fundamental tone plus many overtones) rolling through the universe at once. The different spot sizes tell us the sizes of the waves coursing through the universe. The amount of power in each spot size tells us which spot sizes (i.e., which pressure waves) were dominant. In their analysis of the CMB map, cosmologists have found a pattern with a broad range of wavelengths (i.e., spot sizes) but with a few dominant wavelengths rippling through the young universe. Not all wavelengths and spot sizes were created equal.

The first peak in the CMB power spectrum ($l = 220$) corresponds to an acoustic wave that compressed one time in the first 380,000 years, but did not have enough time to begin expanding. This is the fundamental wave. The second peak ($l = 540$), or first overtone, corresponds to a sound wave that compressed once and then rarefied once in 380,000 years.

These acoustic oscillations in the early universe were set in motion by the disturbance of inflation. Inflation, caused by the release of energy across the entire volume of the universe when the universe was about 10^{-35} seconds old, was not perfectly uniform, as all physical processes will have slight variations, due to quantum fluctuations. Those quantum fluctuations, whose original amplitudes cannot be predicted from theory, were magnified by inflation until they became the sound waves of the post-inflation epoch universe.

The Sound Waves Vanish

For as long as the photons and ionized particles continued to act together like a gas under pressure, the waves generated by inflation were able to continue to propagate through the universe. Those conditions ended abruptly, however, when the universe was 380,000 years old. At that point in time, the photons had been stretched by the expanding fabric of space to wavelengths and energies such that the matter in the universe, coupled in temperature to the sea of photons, was too cool to re-ionize. At that moment, when the electrons and atomic nuclei combined to form and remain as neutral atoms, the free electrons disappeared. Without the free electrons, the photon-dominated gas could no longer support acoustic waves because they had nothing with which they could easily collide. In an instant, the sound waves vanished from the universe and the era of acoustic oscillations ended. As a result, once the universe filled with neutral hydrogen atoms, photons were able to travel, uninhibited by the presence of free electrons, across its entire length and breadth; these photons became the CMB.

The sound waves did, however, leave behind evidence of their presence because when the universe could no longer support pressure waves, the universe was nevertheless speckled with hot (high pressure, high density) and cold (low pressure, low density) spots. As a result, although the acoustic oscillations themselves simply vanished from the universe when the electrons and protons combined and the CMB was released, the sound waves left their ghostly signature, like signs announcing "George Washington slept here" in the form of the hot and cold spots fossilized forever into the CMB pattern.

Regions of the universe corresponding to size scales larger than about one degree (values of $l < 100$ in the power spectrum plot of the CMB) were too large to undergo acoustic oscillations, as the universe was not old enough for the waves to have traveled distances larger than about one degree before the acoustic oscillation era ended. The very large structures in the CMB map, which are seen in the COBE map of the CMB and that show up at small values of l in the power spectrum plot, result from small variations in initial conditions across the universe that were in place before inflation.

The Sizes of the Sound Waves

The CMB map reveals the *angular size* of the strongest temperature fluctuations in the CMB itself, which in turn provides us with a direct measurement of the fundamental frequency of vibration of the gas in the early universe. An angular size, however, is not the same as a *physical size*, but angular size and physical size are related to each other through the physical distance to the object. If we look into the distance and see a skyscraper and measure the angular height of that building as 1 degree, then if we also know the distance to the building (in meters), we can use basic principles of geometry to calculate the actual height of that skyscraper (in meters). Alternatively, if we know the physical height of the building —perhaps we read this information on the architectural blueprints in the archives of the city government's building code office—and the angular size, we can calculate the physical distance to the building.

Conceptually, this technique is analogous to using our measurements of parallax to determine the absolute distance to a star: we measure the parallax angle of the star (equivalent to the angular size of the building). For parallax measurements, we also measure the length of the short side of the triangle (the astronomical unit, which is equivalent to the height of our building). These measurements, along with basic mathematical formulae for triangles, give us all the information we need in order to calculate the length of the long sides of the parallax triangle (the distance to the star or to the building).

Cosmologists know what the contents of the universe were when the acoustic oscillations were in motion. From their knowledge of the physics of plasmas (charged particles) and photons, they can calculate the longest wavelength acoustic wave—143 ± 4 megaparsecs—that the universe could support. This wavelength is the distance that a wave could travel across the universe exactly once (the fundamental) from the Big Bang until the time when the CMB was released; that is, this wavelength was the linear size of the universe at that moment and is the equivalent of our 1 AU length in our parallax analogy. As a result, like reading the information about the skyscraper from the blueprints in the city codes office, we know *a priori* the physical sizes of the acoustic waves. From the patterns in the map of the CMB, cosmologists then measure the angular size—0.601 ± 0.005 degrees (0.0105 ± 0.0001 radians)—of the fundamental mode of vibration in the early universe, which is equivalent to our parallax angle. With measurements in hand for both the physical and angular sizes of these acoustic oscillations, cosmologists can calculate the distance through spacetime from us (today) to the moment when the CMB was released: 44 billion light-years. This number, extracted from information encoded in the WMAP map, tells us something quite fundamental about the universe.

From a Distance to a Time

What does this 44 billion light-years mean?

We know that space has been expanding as the photons travel. Therefore, the photons have not actually crossed a distance of 44 billion light-years; rather, the present distance between us and the location where a particular CMB photon was emitted has been stretched by the expanding universe to become, today, 44 billion light-years in length. We can therefore conclude that the universe must be less than 44 billion years old. But during the time required for this photon to traverse the path from where it was emitted to our telescope, the universe stretched. If we knew by how much it had stretched, we would know the age of the universe. So how old is it?

The distance through expanding spacetime to the CMB surface is closely related, mathematically, to the age of the universe. The age, in turn, depends on how quickly space has been expanding. This relationship therefore must depend on the value of the Hubble constant (which is related to the rate of expansion of the universe after the inflation epoch ended), on the total density of matter in the universe (which slows down the expansion rate), and on the total amount of dark energy in the universe (which accelerates the expansion rate), because together these three parameters control the overall rate of expansion of the universe and the shape of the universe. For example, if the universe were static (neither expanding nor contracting) now and had never changed in size—i.e., if the Hubble constant is now and had always been zero—then the distance that the CMB photons would have traveled through spacetime would tell us that the age of the universe would have to be 44 billion years. On the other hand, if the rate of expansion of the universe was enormous, then the universe could have inflated to its current size almost overnight and the CMB photons would be 44 billion light-years from where they started after the passage of only a few seconds or years, in which case the universe would be in its infancy. So the value of the Hubble constant and the amounts of mass and dark energy in the universe control the relationship between the calculated spacetime distance and the actual age of the universe.

We do have measurements for the value of the Hubble constant in the local (modern) universe and we also have observational data that we have used to determine the relative amounts of dark matter and normal matter. If we could determine the relative amount of dark energy, we would have in our hands all the information we would need to estimate the age of the universe from the WMAP data.

Is Space Flat or Curved?

Cosmologists talk about the total amount of energy (all forms of energy, including all forms of matter plus dark energy) in the universe in terms of the positive or negative curvature or flatness of space. Knowing whether the curvature or the universe is positive or negative or flat is the same as knowing the total amount of matter plus energy that exists in the universe. Do we know whether the universe is flat, in the cosmological sense?

We can get at this concept by thinking about the Earth. Is the surface of the Earth flat or curved? Equivalently, we could ask, Does the Earth have a North Pole, a single point at which all lines of longitude converge? There is a fairly simple test that will decide the matter. Imagine that both you and a friend are standing 100 meters apart on the equator. You use the rising and setting of the Sun to determine the directions east-west and north-south. You each attach one end of an extendable and retractable rope to your respective belt buckles. From your starting positions, you both proceed to walk directly north, ignoring rivers, mountains, oceans, or any other obstacles that might divert either of you from your straight paths. If the Earth is flat, the rope will remain extended to a length of exactly 100 meters and you will never reach the North Pole because there would be no North Pole. If, however, the surface of the Earth is positively curved, as you each walk directly north you will follow independent lines of longitude that bring you closer and closer together and the length of the rope will gradually shorten. When you reach the North Pole, the rope would have retracted completely and you and your friend would bump into each other. On the other hand, if space is negatively curved (imagine both friends walking from the centerline to the front end of the saddle of a very big horse), as you and your friend walk northwards, the rope will stretch out to greater and greater lengths as the separation between the two of you increases. As cosmologist Wayne Hu explains this story, if you and your friend know nothing about space, you might think that some force (the tension of the rope or perhaps gravity) pulled you toward each other as you walked toward the North Pole; however, you walked by following straight lines of longitude on a curved surface. The tension of the rope did not pull you toward each other. What we like to call gravity is nothing more than the positive curvature of space, which is a measure of the amount of mass in the universe.

The curvature of space acts as a lens that focuses light, just as the surface curvature of the Earth focused our two hikers onto paths that converged at the North Pole. As CMB photons travel across vast distances in space, they follow straight lines. If space is positively curved, like the surface of the Earth, those lines come together and the apparent angular sizes of spots, as seen in our maps of the CMB, would appear larger than their actual sizes. More mass, more curvature, larger spots. If the sum of all the mass and energy in the universe is large, the location of all the peaks in the power spectrum plot will be further to the left, toward smaller values of l (bigger spots), whereas if the total amount of mass and energy is small, these peaks will be further to the right (smaller spots), toward larger values of l.

What about the measured sizes of the spots? Where is the location of the first peak in the CMB power spectrum that cosmologists obtained from the CMB map? The first peak is at $l = 220$, which corresponds exactly to the largest spot sizes (0.6 degrees) predicted by cosmologists for a flat universe. We now have the last piece of the puzzle we need in order to tease an age of the universe out of the map of the cosmic microwave background. We just need to put all the pieces together.

A Very Dark Universe

If we live in a flat universe, which is certainly what all the evidence appears to be telling us, then we know the total energy content of the universe. To make the accounting easier, cosmologists discuss the flatness of the universe in terms of a cosmological parameter they call "Omega" (Ω). Ω is the ratio between the actual energy density (energy per cubic meter) of the universe and the energy density that would make the universe flat (cosmologists call this the *critical density*). If the universe has exactly enough energy for the universe to be flat, then $\Omega = 1$. Since we found that $l = 220$ for the first peak in the CMB power spectrum, we know that $\Omega = 1$ is the right answer for our universe.

Remember that the universe contains energy in different forms, so by adding up the energy contained in the forms we understand and know how to measure, we can learn a great deal about the universe. One of these reservoirs of energy is radiation—photons—and virtually all the photons in the universe are in the CMB. After more than 10 billion years, the cosmological redshifting of all of these CMB photons has reduced their collective contribution to the total energy content of universe from about 15 percent at the time the CMB was released to an insignificant amount today.

A second form of energy we understand and know how to measure is normal matter, this being everything in the universe that is made from protons, neutrons, and electrons, including stars, galaxies, clusters of galaxies, and the hot gas between the galaxies. When we add up the energy density of all the photons (almost nothing today) and all the normal matter (about 4.6 percent) in the universe, we find that those two forms of energy contribute less than five percent of the total needed for a flat universe. Since $\Omega = 1$, the other 95 percent of the energy density of the universe must be stored in some other forms. The CMB power spectrum provides information about this 95 percent.

A third form of energy, and a second kind of matter, is the dark matter. Astronomers have abundant evidence to show that it exists, but almost no information about its properties. Dark matter does not produce, emit, or scatter photons. Since the pressure in the early universe was generated almost entirely by photons, the dark matter effectively contributed zero pressure to the young universe and, consequently, did not participate with the normal matter in the acoustic oscillations. As we have seen, however, dark matter does interact gravitationally with normal matter. Since the acoustic waves in the early universe were affected both by pressure (for which dark matter played no role) and gravity (for which dark matter played a significant role), the relative strengths of the different sound waves propagating through the young universe were affected by the amount and the existence of dark matter.

In the power spectrum of the CMB, the odd-numbered peaks record the maximum compression of the gas during the compressive phase of the acoustic oscillations, which is an effect generated by dark and normal matter working together and in the same direction in which the wave is naturally moving. In contrast, the even-numbered peaks correspond to the greatest degree of decompression of the gas during the rarefaction phase; during rarefaction, gas pressure is working against gravity, so the rarefaction waves (e.g., the second and other even-numbered peaks) have less power than the compressive waves (e.g., the first and other odd-numbered peaks). Therefore, the heights of the first and third peaks in the power spectrum relative to the height of the second peak depend sensitively on the amount of normal plus dark matter in the universe. We need also to consider that stronger gravity slows down the acoustic oscillations, so adding dark matter to the universe would decrease the frequencies of these sound waves (i.e., dark matter would shift these oscillations to larger values of l).

Given that the energy densities of radiation and normal matter are measured and total only 4.6 percent of the energy density of the universe today, cosmologists can use the shape of the power spectrum to determine the energy density of dark matter. The answer: about 27.9 percent of the total energy density of the modern universe is in the form of matter, normal plus dark. The difference between these two sums gives us the percentage of the total energy density of the modern universe that is in the form of dark matter: 23.3 percent. In other words, the universe contains about five times more dark matter than normal matter. Reassuringly, this result is consistent with what we have found in the observations of the Bullet Cluster.

Furthermore, if the total energy density of dark matter, normal matter, and radiation add up to just less than 28 percent of the mass and energy content of the universe and if $\Omega = 1$, the other 72.1 percent of the energy density of the universe, our fourth reservoir of energy, must

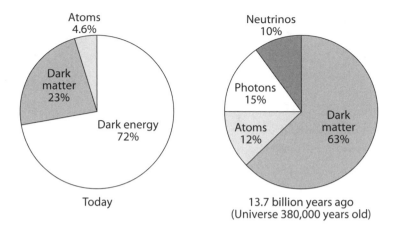

Figure 26.8. Pie charts showing the relative proportions of the principle constituents of the universe today (left) and at the moment the CMB photons were released (right). As the universe has expanded, the density of the dark matter (63 percent down to 23 percent) and of the atoms (12 percent down to 4.6 percent) has decreased, since the number of those particles is essentially fixed while the volume of space has increased. Because photons and neutrinos lose energy as the universe expands, the energy densities of these particles decrease faster than the rate at which the matter density decreases. Together, photons and neutrinos contributed 25 percent of the total energy density of the universe 13.7 billion years ago, but today they contribute almost nothing. The dark energy density does not decrease at all, so over time it becomes more and more prominent as the other principle constituents of the universe decrease in energy. Whereas dark energy made virtually no contribution to the total energy density of the universe 13.7 billion years ago, it now dominates the universe. Image courtesy of NASA and the WMAP science team.

be in the form of dark energy, the energy that powers the accelerating universe (Chapter 21).

The Age of the Universe from WMAP

Now all the puzzle pieces are in place. With $\Omega = 1$, 72.1 percent of the energy content of the universe is in the form of dark energy ($\Omega_{de} = 0.721$), 23.3 percent is in the form of weakly-interacting dark matter ($\Omega_{dm} = 0.233$), and 4.6 percent is in the form of normal matter ($\Omega_{nm} = 0.046$). The distance across expanding space-time to the moment in the history of the universe when the CMB photons were released is 44 billion light-years, and $H = 71.9$ kilometers per second per megaparsec, as derived by the WMAP team from the first five years of data from the WMAP satellite. With the values of all these parameters in hand, cosmologists can calculate the rate at which the universe was expanding just after the CMB photons were released and how that rate has changed with time. That calculation gives us the age of the universe: the age of the universe derived solely from WMAP data is 13.69 billion years. The WMAP team, after including all of their error estimates, and factoring in additional non-WMAP measurements to improve the constraints on the Hubble constant ($H = 70.5$ kilometers per second per megaparsec) calculated an age for the universe of 13.72 billion years.

We should keep in mind that other groups, using supernova measurements from the Hubble Space Telescope, have recently calculated a slightly larger value for the Hubble constant. The revised value would drop the estimated WMAP age of the universe to a bit below 13.7 billion years. Meanwhile, planned NASA missions will improve our measurements of the dark energy content of the universe, and additional WMAP measurements will continue to refine the answers derived from the WMAP data. So while we cannot definitively say, from the WMAP results, that the universe is 13.69 or 13.72 billion years old, we can say that the WMAP results are remarkably robust in indicating an age in the vicinity of 13.5 to 14 billion years. And we can take considerable comfort from the realization that this age estimate is in impressive agreement with the estimates reached by other methods of calculating the age of the universe.

CHAPTER 27

A Consistent Answer

How old is the universe? For several centuries, astronomers have been making observations of the nighttime sky, and gradually we have made progress toward solving the riddles of the heavens. By combining observations with the laws of physics, we have teased enormous amounts of information from the light we collect with our telescopes. From this accumulated effort of a dozen generations of scientists, what have we learned about the age of the universe?

Logically, we know that the universe must be older than the Earth, the Moon, and our solar system. We know that the oldest rocks on all the Earth's continents are between 3.6 and 4.0 billion years old while the oldest mineral grains on Earth, found in sedimentary rocks, are 4.3 to 4.4 billion years old. The Earth, clearly, must be at least this old. Lunar rocks, brought back to Earth by the Apollo 15, 16, and 17 astronauts, have radiometric ages of 4.4 to 4.5 billion years. The oldest meteorites in our solar system, which may be among the very first objects that formed in our solar system, are 4.567 billion years old. From all of our knowledge about how stars, planets, and the parent bodies of meteorites form, we are confident that the Sun, Moon, Earth, and the other planets were all born at nearly the same time. The Earth and Sun therefore are very nearly 4.5 billion years old. This age for the Sun is consistent with all of our observations and our theoretical understanding of the laws of physics that control the size, temperature and luminosity of the Sun and of other stars, and with our knowledge of how stars change with time as a result of nuclear fusion. The universe must, of course, be older than its contents, so we can assign to it a minimum age of 4.6 billion years old. But we can do better than that.

At the end of their lifetimes as normal stars, stars become black holes, neutron stars, or white dwarfs. Isolated white dwarfs, these being stars outside of binary or multiple star systems, have no sources of energy and cannot change in mass. Since they are hot, they radiate heat into space. Their sizes, however, are completely determined by the pressure produced inside them by degenerate electrons, which means, quite simply, that they do not get smaller in diameter no matter how cool they become. As a result, white dwarfs radiate heat into space and cool off, but they do not shrink. From the basic physics of white dwarfs, we can calculate the rate at which a white dwarf will cool off; we can also look into the past and calculate how much warmer today's white dwarf was yesterday, last year, 1 million years ago, or 10 billion years ago. The temperatures of the coolest and oldest white-dwarf stars in the Milky Way Galaxy today allow us to calculate backwards in time to determine that they have been cooling for at least 11 billion years and perhaps as much as 13 billion years. By comparison, the Earth and our solar system are youngsters. Since the galaxy had to form first, and then some stars had to form, live, and die before they could begin to cool as white dwarfs, the age of the universe must be at least a little bit greater than 11 to 13 billion years. From this result, we do not yet know the age of the universe, but we can say with confidence that if the universe were significantly older than 13 billion years, we would find white dwarfs that are substantially cooler than the coolest ones we have found. If no such supercool white dwarfs exist—and it appears that they do not—the universe is most likely only a little bit older than 13 billion years. What else do we know?

We know from studying stars that the most massive stars are also the most luminous stars. This relationship between mass and luminosity is a consequence of the force of gravity: more mass in the same volume of space generates a stronger compressive force. More massive stars squeeze themselves harder. To put the brakes on gravity squashing stars into infinitely small volumes, stars must find a way to push back, and they do this by generating heat from nuclear fusion reactions. The most massive stars generate greater inward gravitational forces than less massive stars, so they must generate more thermal pressure than lower-mass stars in order to resist gravity. Greater thermal pressure comes from more rapid rates of nuclear reactions in the cores of these stars, which release energy faster and thereby maintain the high internal temperatures needed to resist gravity as well as the high luminosities seen from their surfaces. Consequently, massive stars use up their nuclear fuel more quickly than less massive stars. In other words, because less massive stars are less profligate with their fuel, they can live longer. This correlation between mass, luminosity, and the maximum lifetime of a star leads directly to another method for determining the age of the universe, which involves the turn-off point identified on an H-R diagram for a globular star cluster. The oldest globular clusters in the Milky Way have ages of about 13 billion years. The Milky Way and the universe itself must be at least a few hundred million years older than these oldest globular clusters, based on the amount of the element beryllium that the latter contain. Reassuringly, this age is very consistent with the age obtained from the cooling ages of white dwarfs. Do we know anything else?

We know that the universe is expanding, and the rate at which the universe is expanding provides another technique by which astronomers may estimate the age of the universe. All galaxies outside of our immediate neighborhood appear to be moving away from us. This apparent motion of the galaxies, which we observe through the redshift of the light from stars in those galaxies, is a consequence of the expansion of the universe. Astronomers have now detected Cepheid variable stars in galaxies out to distances of 30 megaparsecs (100 million light-years). The brightnesses of these Cepheids permit us to calculate directly the distances to the host galaxies of the Cepheids. These distances, combined with the redshift velocities of the host galaxies, allow us to measure the value of the Hubble constant. Assuming that the expansion rate of space has been constant over the history of the universe, which all evidence suggests is very close to a correct assumption, the most accurate measurements of the value of the Hubble constant yield an age of the universe of 13.5 billion years, with a margin of error of about 2 billion years. When we include all our knowledge about dark matter and dark energy, we recognize that the expansion rate of the space has not been perfectly constant for the last 13 billion years, and we get an age estimate for the universe of about 13 billion years, with the same 2 billion year margin of error. We could cautiously con-

TABLE 27.1.
Scientific Measurements for the Age of the Universe

Technique	Derived Age of the Universe
Cooling times for white dwarfs in Milky Way	> 11–13 billion years
Absence of extremely cool, faint white dwarfs	< 15 billion years
Turn off-points for globular star clusters	13–14 billion years
Expansion rate of the universe	13.5–14 billion years
Cosmic background radiation power spectrum analysis	13.5–14 billion years

clude that the Hubble constant-derived age of the universe is in the range from 11 to 15.5 billion years, which is consistent with the answer obtained from white-dwarf cooling times and turn-off points for stars in globular clusters. Now we have three different, completely independent methods that place the age of the universe at about or a little more than 13 billion years, with strong evidence that an age of 13.5 to 14 billion years is very likely.

Finally, cosmologists have teased an estimate for the age of the universe out of our best maps of the cosmic microwave background, the remnant signal of the Big Bang. To do this, they must know the total energy content of the universe, which they obtain from measurements of the flatness of the universe, the current and past expansion rates of the universe, and the total amount of dark energy, dark matter, and normal matter. From all of these measurements, cosmologists calculate that the age of the universe is in the vicinity of 13.7 billion years. From these results, we can state with reasonable confidence that the age of the universe is likely between 13.5 and 14 billion years.

Now we have four independent methods for deriving an age for the universe, and all four yield a consistent answer. If the age derived by any one of these methods were presented by itself, we might reasonably wonder whether to believe the claim. Taken together, however, we now have extremely strong evidence that the universe is indeed 13.5 to 14 billion years of age. This whole Big Shebang got started just a bit over 13.5 billion years ago. The trials and errors, painstaking observations and brilliant insights that have led to this answer amount to one of mankind's the most impressive intellectual accomplishments.

∴ INDEX ∴